Teachers Engaged in Research

Inquiry Into Mathematics Classrooms, Grades 6–8

a volume in
Teachers Engaged in Research

Series Editor
Denise S. Mewborn
University of Georgia

Library of Congress Cataloging-in-Publication Data

Teachers engaged in research : inquiry into mathematics classrooms, grades 6-8 / edited by Joanna O. Masingila.
    p. cm. – (Teachers engaged in research)
  Includes bibliographical references.
  ISBN 1-59311-500-8 (hardcover) – ISBN 1-59311-499-0 (pbk.)
  1. Mathematics–Study and teaching (Middle school) 2. Effective teaching.
3. Middle school teaching–Research. I. Title: Inquiry into mathematics classrooms, grades 6-8. II. Masingila, Joanna O. (Joanna Osborne), 1960-
III. Series.
  QA11.2.T423 2006
  510.71'2–dc22

                        2006007072

Printed in the United States of America

Teachers Engaged in Research

Inquiry Into Mathematics Classrooms, Grades 6–8

edited by

Joanna O. Masingila

Syracuse University

INFORMATION AGE
PUBLISHING

Greenwich, Connecticut • www.infoagepub.com

ACKNOWLEDGMENTS

Producing a series like this requires the diligent work of many people, and that is certainly the case with this volume. I am grateful to the authors of the chapters in this volume for their tireless efforts in producing high-quality chapters representing their research work. I wish to thank the reviewers—Amy Alpen, Shuhua An, Manja Ericsson, Marian Fox, Doug Franks, Dana Griffith, Jean Hallagan, Mary Koster, Catie Lambie, Michael Lutz, Karen Madden, Jon Manon, Doug McDougall, Kathleen Morris, Christina Pfister, Michele Salamy, Christy Snook, Lynn Stallings, Victoria Waldron and Linda Wilson—who provided scholarly advice and detailed comments that resulted in helpful feedback to the authors. I am also grateful to Elizabeth DePriest, a mathematics education graduate student at Syracuse University, for assisting with formatting the chapters.

I am especially indebted to my colleagues—Denise Mewborn, series editor, Stephanie Smith and Marvin Smith, grades P–2 volume editors, Cindy Langrall, grades 3–5 volume editor, Laura Van Zoest, grades 9–12 volume editor, Gini Stimpson, consultant, and Bea D'Ambrosio, consultant—for thought-provoking meetings and lively discussions, in person and via e-mail. Giving voice to classroom teachers' research is extremely important and the work was quite enjoyable as a collaborative effort.

Joanna O. Masingila, Grades 6–8 Volume Editor
Syracuse, New York

LIST OF CONTRIBUTORS

Laurinda Brown	University of Bristol, Bristol, UK
Cynthia H. Callard	Twelve Corners Middle School/ University of Rochester, Rochester, NY
Marilyn Cochran-Smith	Boston College, Boston, MA
Alf Coles	Kingsfield School, South Gloucestershire, UK
Beatriz D'Ambrosio	Miami University, Miami, OH
Marcia DeJesús-Rueff	Expeditionary Learning Outward Bound, Rochester, NY
Amy French	Angevine Middle School, Lafayette, CO
Maureen Grant	Metropolitan School District of Washington Township, Indianapolis, IN
Eric Gutstein	University of Illinois-Chicago, Chicago, IL
Shelly Sheats Harkness	University of Cincinnati, Cincinnati, OH
Annette Hill	South Dade Senior High School, Homestead, FL
Signe Kastberg	Indiana University-Purdue University at Indianapolis, Indianapolis, IN
Judith Koenig	Retired from Boulder Valley School District, Boulder, CO
Art Mabbott	Seattle Public Schools, Seattle, WA
Joanna O. Masingila	Syracuse University, Syracuse, NY
Mitchell J. Nathan	University of Wisconsin-Madison, Madison, WI
K. Ann Renninger	Swarthmore College, Swarthmore, PA
Carmelita A. Santiago	Atlanta Public Schools, Atlanta, GA
Mary C. Shafer	Northern Illinois University, De Kalb, IL
Tracey J. Smith	Charles Sturt University, Wagga Wagga, New South Wales, Australia
Susan Stein	Portland State University, Portland, OR
Christine D. Thomas	Georgia State University, Atlanta, GA
Vicki Walker	Indiana University-Purdue University at Indianapolis, Indianapolis, IN

CONTENTS

SERIES FOREWORD

Marilyn Cochran-Smith
Boston College

This series, *Teachers Engaged in Research: Inquiry Into Mathematics Classrooms*, represents a remarkable accomplishment. In four books, one each devoted to teaching and learning mathematics at different grade level groupings (Pre-K–2, 3–5, 6–8, and 9–12), ninety-some authors and co-authors write about their work as professional mathematics educators. Across grade levels, topics, professional development contexts, schools, school districts, and even nations, the chapters in these four books attest to the enormous complexity of teaching mathematics well and to the power of inquiry as a way of understanding and managing that complexity.

In a certain sense, of course, the education community already knows that doing a good job of teaching mathematics is demanding—and all too infrequent—and that it requires deep and multiple layers of knowledge about subject matter, learners, pedagogy, and contexts. Nearly two decades of ground-breaking research in cognition and mathematics education has told us a great deal about this. But the books in this series are quite different from most of what has come before. These four books tell us about teaching and learning mathematics from the inside—from the perspectives of school-based teacher researchers who have carefully studied the commonplaces of mathematics teaching and learning, such as the whole

Teachers Engaged in Research
Inquiry Into Mathematics Classrooms, Grades 6–8, pages ix–xx
Copyright © 2006 by Information Age Publishing

class lesson, the small group activity, the math problem, the worksheet where the student shows his or her work, the class discussion about solutions and answers, and the teacher development activity. Taking these and other commonplaces of mathematics teaching and learning as sites for inquiry, the multiple authors of the chapters in these four books offer richly textured and refreshingly insightful insider accounts of mathematics teaching and learning. Reflecting on the contrasts that sometimes exist between teachers' intentions and the realities of classroom life, the chapters depict teachers in the process of considering and reconsidering the strategies and materials they use. Drawing on students' work and classroom discourse, the chapters show us what it looks like as teachers strive to make sense of and capitalize on their students' reasoning processes, even when they don't end up with traditional right answers. Staying close to the data of practice, the chapters raise old and new questions about how students—and teachers—learn to think and work mathematically. In short, this remarkable quartet of books reveals what it really means—over time and in the context of different classrooms and schools—for mathematics teachers to engage in inquiry. The goal of all this inquiry is nothing short of a culture shift in the teaching of mathematics—the creation of classroom learning environments where the focus is deep understanding of mathematical concepts, practical application of skills, and problem solving.

These four books provide much-needed and rich detail from the inside about the particulars of mathematics teaching and learning across a remarkably broad range of contexts, grade levels, and curricular areas. In addition, the books tell us a great deal about teachers' and pupils' learning over time (and the reciprocal relationships of the two) as well as the social, intellectual and organizational contexts that support their learning. Undoubtedly teachers and teacher educators will find these books illuminating and will readily see how the authors' successes and the struggles resonate with their own. If this were the only contribution this series made, it would be an important and worthwhile effort. But this series does much more. The books in this series also make an important contribution to the broader education community. Taken together as a whole, the chapters in these four books have the potential to inform mathematics education research, practice and policy in ways that reach far beyond the walls of the classrooms where the work was originally done.

TAKING AN INQUIRY STANCE

This series of books shows brilliantly and in rich and vivid detail what it means for teachers to take an inquiry stance on teaching and learning mathematics in K-12 classrooms over the professional lifespan. "Inquiry stance" is a concept that my colleague, Susan Lytle, and I have written about over the last decade and a half. It grew out of the dialectic of our simultaneous work as teacher educators involved in day-to-day, year-to-year participation in teacher learning communities, on the one hand, and as researchers engaged in theorizing the relationships of inquiry, knowledge, and practice, on the other. In our descriptions of inquiry as stance, we have repeatedly emphasized its distinction from "inquiry as project." By inquiry project (as opposed to stance), we refer to things like the classroom study that is the required culminating activity during the student teaching semester in a preservice program where inquiry is not integral and not infused throughout, or, the day-long workshop on teacher research or action research that is part of a catalog of professional development options for experienced teachers rather than a coherent part of teacher learning over the lifespan.

Teacher research or inquiry *projects* are discrete and bounded activities, often carried out on a one-time or occasional basis. In contrast, as we have suggested (Cochran-Smith & Lytle, 1999), when inquiry is a *stance*, it extends across the entire professional lifespan, and it represents a world view or a way of knowing about teaching and learning rather than a professional project:

> In everyday language, "stance" is used to describe body postures, particularly with regard to the position of the feet, as in sports or dance, and also to describe political positions, particularly their consistency (or the lack thereof) over time. In the discourse of qualitative research, "stance" is used to make visible and problematic the various perspectives through which researchers frame their questions, observations, and interpretations of data. In our work, we offer the term *inquiry as stance* to describe the positions teachers and others who work together in inquiry communities take toward knowledge and its relationships to practice. We use the metaphor of stance to suggest both orientational and positional ideas, to carry allusions to the physical placing of the body as well as to intellectual activities and perspectives over time. In this sense, the metaphor is intended to capture the ways we stand, the ways we see, and the lenses we see through. Teaching is a complex activity that occurs within webs of social, historical, cultural and political significance. Across the life span an inquiry stance provides a kind of grounding within the changing cultures of school reform and competing political agendas.

> *Inquiry as stance* is distinct from the more common notion of inquiry as time-bounded project or activity within a teacher education course or professional development workshop. Taking an inquiry stance

means teachers and student teachers working within inquiry communities to generate local knowledge, envision and theorize their practice, and interpret and interrogate the theory and research of others. Fundamental to this notion is the idea that the work of inquiry communities is both social and political—that is, it involves making problematic the current arrangements of schooling, the ways knowledge is constructed, evaluated, and used, and teachers' individual and collective roles in bringing about change. To use *inquiry as stance* as a construct for understanding teacher learning in communities, we believe that we need a richer conception of knowledge than that allowed by the traditional formal knowledge-practical knowledge distinction, a richer conception of practice than that suggested in the aphorism that practice is practical, a richer conception of learning across the professional life span than that implied by concepts of expertise that differentiate expert teachers from novices, and a rich conception of the cultures of communities as connected to larger educational purposes and contexts . (pp. 288–289)

Developing and sustaining an inquiry stance on teaching, learning and schooling is life-long and constant pursuit for beginning teachers, experienced teachers, and teacher educators alike. A central aspect of taking an inquiry stance is recognizing that learning to teach is a process that is never completed but is instead an ongoing endeavor. The bottom-line purpose of taking an inquiry stance, however, is not teachers' development for its own sake (although this is an important and valuable goal), but teachers' learning in the service of enriched learning opportunities for Pre-K–12 students.

The chapters in these four books vividly explicate the work of Pre-K–12 teacher researchers who engage in inquiry to transform their teaching practice, expand the mathematical knowledge and skill of their students, and ultimately to enhance their students' life chances in the world. Across the 50 chapters in these books, the serious intellectual work of mathematics teaching is revealed again and again as are the passion teachers have for their work (and their students) and the complexity of the knowledge teachers must have to support students' emerging mathematical development.

INQUIRY INTO TEACHING AND LEARNING IN MATHEMATICS

Over the last two decades, various forms of practitioner inquiry, such as teacher research, action research, and collaborative inquiry, have become commonplace in preservice teacher preparation programs as well as in professional development projects and school reform efforts of various kinds. Although there are many variations and often different underlying assumptions among these, the use of inquiry has generally reflected a move away from transmission models of teacher training and retraining wherein teachers are expected to implement specific practices developed by others. The move has been toward a concept of teacher learning as a life-long process of both posing and answering questions appropriate to local contexts, working within learning communities to construct and solve problems, and making all of the aspects of the work of teaching possible sites for inquiry.

As this series of books demonstrates so clearly, an inquiry stances is quite compatible with the professional standards for teaching and learning mathematics that have emerged since the mid 1980s. In fact, in a number of the chapters in these books, the explicit purpose of the researchers is to study the implementation and effectiveness of standards-driven mathematics teaching and learning. The chapters concentrate on a wide range of mathematics topics, for example: concepts of number, addition and subtraction, algebraic thinking, linear measurement, geometric patterns and shapes, classification and sorting, mathematical proofs, multiplicative reasoning, division of fractions, data analysis and probability, volume, and the concept of limit. The chapters also examine what happens in terms of students' learning and classroom culture when various strategies and teaching methods are introduced: bringing in new models and representations of mathematical concepts and operations, teaching with problems, using writing in mathematics instruction, having students share their solution strategies, supporting students' own development of the algorithms for various mathematical operations, encouraging students to work in small groups and other collaborative arrangements, including role play in one's repertoire of mathematics teaching strategies, and integrating multiple opportunities for students to participate in hands-on technology and to work with technology-rich problems. Taken as a group, the chapters in these four volumes illustrate the power of an inquiry stance to widen and deepen teachers' knowledge of the subject matter of mathematics at the same time that this stance also enhances teachers' understanding of the learning processes of their students. Both of these—deeper knowledge of mathematics and richer understandings of learners and learning—are essential ingredients in teaching mathematics consistent with today's high standards.

Some of the chapters in these four volumes are written by single authors. These chronicle the ongoing efforts of teachers to understand what is going on in their classrooms so they can build on the knowledge and experiences students bring at the same time that they expand students' knowledge and skill. It is not at all surprising, however, that many of the chapters are co-authored, and even the single-authored chapters often reflect teachers' experiences as part of larger learning communities. Inquiry and community go hand in hand. The inquiries reported in these volumes feature collaborations among teachers and a whole array of colleagues, including fellow teachers, teacher study groups, teacher educators, university-based researchers, professional development facilitators, curriculum materials developers, research and development center researchers, and directors of large-scale professional development projects.

Across the chapters, there is a fascinating range of collaborative relationships between and among the teacher researchers and the subjects/objects of their inquiries: two teachers who collaborate to expand their own knowledge of mathematical content, the teacher researcher who is the subject of another researcher's study, the school-based teacher researcher and the university-based researcher who form a research partnership, the teacher who collaborates with a former teacher currently engaged in graduate study, pairs and small groups of teachers who work together to implement school-wide change in how mathematics is taught, teachers who team up to examine whether and how standards for mathematics teaching and learning are being implemented in their classrooms, teachers who engage in inquiry as part of their participation in large-scale professional development projects, the teacher who is part of a team that developed a set of professional development materials, a pair of teachers who collaborate with a university-based research and development group, a group of teachers who develop an inquiry process as a way to induct new teachers into their school, two teachers who inquire together about pedagogy and practice in one teacher's classroom, and teachers who form research groups and partnerships to work with teacher educators and student teachers. Almost by definition, practitioner inquiry is a collegial process that both occurs within, and stems from, the collaborations of learning communities. What all of these collaborations have in common is the assumption that teacher learning is an across-the-professional-lifespan process that is never "finished" even though teachers have years of experience. The chapters make clear that learning from and about teaching through inquiry is important for beginning and experienced teachers, and the intellectual work required to teach mathematics well is ongoing.

The inquiries that are described in these four books attest loudly and clearly to the power of questions in teachers' and students' learning. These inquiries defy the norm that is common to some schools where the competent teacher is assumed to be self-sufficient and certain and where asking questions is considered inappropriate for all but the most inexperienced teachers. Similarly, the chapters in these four books challenge the myth that good teachers rarely have questions they cannot answer about their own practices or about their students' learning. To the contrary, although the teachers in these four books are without question competent and may indeed be self-sufficient and sometimes certain, they are remarkable for their questions. Teachers who are researchers continuously pose problems, identify discrepancies between theory and practice, and challenge common routines. They continuously ask critical questions about teaching and learning and they do not flinch from self-critical reflection: Are the students really understanding what the teacher is teaching? What is the right move to make at various points in time to foster students' learning? How does the teacher know this? Are new curricular materials and teaching strategies actually supporting students' learning? How do students' mathematical ideas change over time? What is the evidence of students' growth and development? How can theory guide practice? How can practice add to, even alter, theory? These teachers ask questions not because they are failing, but because they—and their students—are learning. They count on their collaborators for alternative perspectives about their work and alternative interpretations of what is going on. In researching and writing about their work, they make explicit and visible to others both the decisions that they make on an ongoing basis and the intellectual processes that are the backdrop for those decisions. Going public with questions, seeking help from colleagues, and opening up one's practice to the scrutiny of outsiders may well go against the norms of appropriate teaching behaviors in some schools and school districts. But, as the chapters in this series of books make exquisitely clear, these are the very activities that lead to enriched learning opportunities and expanded knowledge for both students and teachers.

In this current era of accountability, there is heavy emphasis on evidence-based education and great faith in the power of data and research to improve educational practice. The chapters in the four books in this series illustrate what it looks like when mathematics teaching and learning are informed by research and evidence. In some of the chapters, for example, teacher researchers explicitly examine what happens when they attempt to implement research-based theories and teaching strategies into classroom practice. In these instances, as in others, practice is driven by research and evidence in several ways. Teacher researchers study the work of other researchers, treating this work as generative and illuminating, rather than regarding it as prescriptive and limiting. They reflect continuously about how others' research can (and should) inform their curricular, instructional, and assessment decisions about teaching mathematics. They also examine what consequences these decisions have for students' learning by collecting and analyzing a wide array of classroom data—from young children's visual representations of mathematical concepts and operations to classroom discussions about students' differing solutions to math problems to pre- and post-tests of students' mathematical knowledge and skill to interviews with students about their understandings to students' scores on standardized achievement tests.

In many of the chapters, teachers explicitly ask what it is that students know, what evidence there is that students have this knowledge and that it is growing and developing, and how this evidence can be used to guide their decisions about what to do next, with whom, and under what conditions. Braided together with these lines of inquiry about students' learning are questions about teachers' own ways of thinking about mathematics knowledge and skill. It is interesting that in many of the inquiries described in these four books, it is difficult to separate process from product or to sort out instruction from assessment. In fact, many of the chapters reveal that the distinctions often made between instruction and assessment and between process and product are false dichotomies. When mathematics teaching is guided by an inquiry stance and when teachers make decisions based on the data of practice, assessment of students' knowledge and skill is ongoing and is embedded into instruction, and the products of students' learning are inseparable from their learning and reasoning processes.

Finally, in some of the chapters in this quartet of books, teacher researchers and their colleagues use the processes of inquiry to examine issues of equity and social justice in mathematics teaching and learning. For example, one teacher problematizes commonly used phrases and ideas in mathematics education such as "success," particularly for those who have been previously unsuccessful in mathematics. Another explores connections between culture and mathematics by interviewing students. A trio of teachers examines the mathematical understandings of their students with disabilities, while another teacher chronicles his efforts to construct mathematics problems and projects that focus on social justice. Each of these educators values and draws heavily on the data of students' own voices and perspectives; each works to empower students as active agents in their own learning. These examples are a very important part of the collection of inquiries in this book series. When scholars write about teaching and teacher education for social justice, mathematics is the subject matter area least often included in the discussion. When teachers share examples of teaching for social justice in their own schools and classrooms, mathematics is often the area they have the most difficulty incorporating. When student teachers plan lessons and units related to equity and diversity, mathematics is often the subject area for which they cannot imagine any connection. The chapters in these books that specifically focus on equity and social justice make it clear that these issues readily apply to mathematics teaching and learning. But even in the many chapters for which these issues are not the explicit focus, it is clear that the teachers' intention is to empower all students—even (and especially) those groups least well served by the current educational system—with greater mathematical acuity and agency.

THE VALUE OF TEACHERS' INQUIRIES INTO
MATHEMATICS TEACHING AND LEARNING

There is no question that engaging in inquiry about teaching and learning mathematics is an important (and often an extremely powerful) form of professional development that enhances teachers' knowledge, skill and understandings. This is clear in chapter after chapter in the four volumes of this series. The authors themselves describe the process of engaging in self-critical systematic inquiry as transformative and professionally life-changing. They persuasively document how their classroom practices and their ways of thinking about mathematical knowing changed over time. They combine rich and multiple data sources to demonstrate growth in their students' knowledge and skill. Undoubtedly many Pre-K–12 teachers and school- and university-based teacher educators will find the inquiries collected in these four volumes extraordinarily helpful in furthering their own work. The questions raised by the teacher researchers (often in collaboration with university-based colleagues and others) will resonate deeply with the questions and issues of other educators striving to teach mathematics well by fostering conceptual understandings of big mathematical ideas, reliable but imaginative problem solving strategies, and solid mathematical know-how about practical applications to everyday problems. The teacher researchers whose work is represented in these volumes have made their questions and uncertainties about mathematics teaching explicit and public. In doing so, they have offered their own learning as grist for the learning and development of other teachers, teacher educators, and researchers. This is a major contribution of the series.

Few readers of the volumes in this series will debate the conclusion that systematically researching one's own work as a mathematics educator is a valuable activity for teacher researchers themselves, for their students in Pre-K–12 schools and classrooms, and for other Pre-K–12 teachers and teacher educators who are interested in the same issues. But some readers will raise questions about whether or not there is value to this kind of inquiry that carries beyond the participants involved in the local context or that extends outside the Pre-K–12 teaching/teacher education community. After all, the argument of some skeptics might go, for the most part the inquiries about mathematics teaching and learning that are included in this series were prompted by the questions of individual teachers or by small local groups of practitioners working in collaboration. With a few exceptions, these inquiries were conducted in the context of a single classroom, course,

school, or program. Almost by definition then, the skeptics might say, few of
the inquiries in these four volumes can hold up to traditional research criteria
for transferability and application of findings to other populations and contexts (especially if conceptualized as the identification of causes and
effects). For this reason, the skeptics might conclude, the kind of work
included in this series is nothing more than good professional development for individuals in local communities, which may be of interest to
other teacher educators and professional developers.

Putting aside for the moment that good professional development is
essential in mathematics education (not to mention, hard to come by), the
skeptics' line of argument, as outlined in the paragraph above, is both
short-sighted and uninformed. The inquiries that are part of the *Teachers
Engaged in Research: Inquiry Into Mathematics Classrooms* series are valuable
and valid well beyond the borders of the local communities where the work
was done, and the criteria for evaluating traditional research are not so
appropriate here. With many forms of practitioner inquiry, appropriate
conceptions of value and validity are more akin to the idea of "trustworthiness" that has been forwarded by some scholars as a way to evaluate the
results of qualitative research. From Mishler's (1990) perspective, for
example, the concept of validation ought to replace the notion of validity.
Validation is the extent to which a particular research community, which
works from "tacit understandings of actual, situated practices of a field of
inquiry" (Lyons and LaBoskey, 2002, p. 19), can rely on the concepts,
methods, and inferences of an inquiry for their own theoretical and conceptual work. Following this line of reasoning, validation rests on concrete examples (or "exemplars") of actual practices presented in enough
detail so that the relevant community of practitioner researchers can
judge the trustworthiness and usefulness of the observations and analyses
of an inquiry.

Many of the inquiries in these four books offer what may be thought of
as exemplars of teaching and learning mathematics that are consistent
with current professional standards in mathematics education. Part of what
distinguishes the inquiries of the teacher researchers in these four books
from those of outside researchers who rely on similar forms of data collection is that in addition to documenting students' learning, teacher
researchers have the opportunity and the access to systematically document their own teaching and learning as well. They can document their
own thinking, planning, and evaluation processes as well as their questions, interpretive frameworks, changes in views over time, issues they see
as dilemmas, and themes that recur. Teacher researchers are able to analyze the links as well as the disconnects between teaching and learning that
are not readily accessible to researchers based outside the contexts of practice. Systematic examination and analysis of students' learning juxtaposed

and interwoven with systematic examination of the practitioners' intentions, reactions, decisions, and interpretations make for incredibly detailed and complex analyses—or exemplars—of mathematics teaching and learning.

With rich examples, the teacher researchers in these books demonstrate how they came to understand their students' reasoning processes and thus learned to intervene more adeptly with the right question, the right comment, a new problem, or silent acknowledgement and support. Braiding insightful reflections on pedagogy with perceptive analyses of their students' understandings, the authors in these four books make explicit and visible the kinds of thinking and decision making that are usually implicit and invisible in studies of mathematics teaching. Although they are usually invisible, however, these ways of thinking and making teaching decisions are essential ingredients for teaching mathematics to today's high standards. I believe that a critical contribution of this series of books—a contribution that extends well beyond the obvious benefits to the participants themselves—is the set of exemplars that cuts across grade levels and mathematics topics. What makes this series unique is that the exemplars in this book do not emerge from the work of university-based researchers who have used the classroom as a site for research and for the demonstration of theory-based pedagogy. Rather the trustworthiness—or validation—of the exemplars in these four books derives from the fact that they were conducted by school-based teacher researchers who shouldered the full responsibilities of classroom teaching while also striving to construct appropriate curriculum, develop rich teaching problems and strategies, and theorize their practice by connecting it to larger philosophical, curricular, and pedagogical ideas.

This series of books makes a remarkable contribution to what we know about mathematics teaching and learning and about the processes of learning to teach over time. Readers will be thoroughly engaged.

REFERENCES

Cochran-Smith, M., & Lytle, S. L. (1999). Relationship of knowledge and practice: Teacher learning in communities. In A. Iran-Nejad & C. Pearson (Eds.), *Review of research in education* (Vol. 24, pp. 249–306). Washington, DC: American Educational Research Association.

Lyons, N., & LaBoskey, V. K. (Eds.). (2002). *Narrative inquiry in practice: Advancing the knowledge of teaching.* New York: Teachers College Press.

Mishler, E. (1990). Validation in inquiry-guided research: The role of exemplars in narrative studies. *Harvard Educational Review, 60*(4), 415–442.

CHAPTER 1

INTRODUCTION TO
THE 6–8 VOLUME

Joanna O. Masingila[1]
Syracuse University

What is a teacher's role in research? Certainly teachers have been the *impetus* of research studies throughout the history of educational research. They have also been referred to as *consumers* of research. It is unusual, however, for teachers to be considered *producers* of research. Yet it is not unusual for teachers to be engaged in inquiry. In fact, some would consider this a fair description of what it means to teach. The difference is that for most teachers, the process as well as the product of their inquiry is tacit. It may not be well defined in terms of specific research questions or systematic in terms of data collection and analysis. Furthermore, their work may not be presented to colleagues for discussion and review or disseminated for publication. Those scholarly activities have typically not been part of the culture of teaching in most school systems.

That may be changing, as demonstrated by the chapters in this volume written by grade 6–8 teachers from Australia, England, and across the United States. Teachers are engaging in research in a variety of ways and their voices are gaining presence in the mathematics education research

---

Teachers Engaged in Research
Inquiry Into Mathematics Classrooms, Grades 6–8, pages 1–14
Copyright © 2006 by Information Age Publishing

community. The authors who contributed to this volume were involved in research through a variety of activities, including:

- reading and reflecting on a variety of research and other literature in the field;
- interpreting findings from the research literature to influence their instructional practice;
- participating in study groups with their colleagues;
- generating research questions for themselves and others to investigate;
- participating in research studies and professional development projects led by other researchers; and
- designing and implementing their own studies and sharing their findings.

It is clear from these chapters that the authors' involvement in these research activities had a profound impact on their conceptions of teaching and learning. Their teaching changed as a result of their inquiries and thus these research activities had an immediate, direct impact on practice. Through the sharing of their stories, we hope to broaden that impact beyond their classrooms and schools to the wider mathematics education community.

Through these chapters, we get a glimpse into the questions that capture the attention of teachers, the methodologies that they use to gather data, and the ways in which they make sense of what they find. Some of the research findings are preliminary, tentative; others are confirmatory; and some are groundbreaking. In all cases, they provide fodder for further thinking and discussion about critical aspects of mathematics education.

HOW THIS VOLUME CAME TO BE

History of the Series

The idea for the *Teachers Engaged in Research* series arose from efforts by the Research Advisory Committee (RAC) of the National Council of Teachers of Mathematics (NCTM) to expand traditional conceptions of research in mathematics to include practitioner inquiry and the research questions that are of interest to practitioners. Beginning in the late 1990s, the RAC made a concerted effort to recognize teachers as producers (and not just consumers) of research. To this end, NCTM sponsored a Working Conference on Teacher Research in Mathematics Education in Albuquerque, NM in 2001. The goal of this conference was to articulate a list of issues that should be considered in developing a framework for teacher research in mathematics education. This conference led to a grant proposal that would have brought together participants from the conference for a writing workshop with the goal of producing a publication similar to this one. For a variety of reasons, that project did not come to fruition. The RAC then turned its attention to including teacher researchers as participants—both attendees and speakers—in the Research Presession that precedes the annual NCTM meeting.

During this time, there was ongoing dialogue in the RAC about a publication that would highlight practitioner inquiry. Simultaneously, the RAC and the Educational Materials Committee (EMC) decided that it was time to issue an update of the *Handbook of Research on Mathematics Teaching and Learning* (Grouws, 1992). Unbeknownst to many, the series titled *Research into Practice* edited by Sigrid Wagner was originally conceived by the EMC as a "companion" to the *Handbook*. The *Research into Practice* series featured chapters co-authored by university-based researchers and classroom teachers and reflected a view of teachers as consumers of research—an accurate reflection of the field at the time. When the RAC and EMC discussed the update of the *Handbook* and its companion, the companion was recast to reflect a view of teachers as producers of research. The *Handbook* (Lester, in preparation) documents a portion of the knowledge base available in mathematics education; the four volumes of the *Teachers Engaged in Research* series fill a gap in that knowledge base by focusing on the research contributions of classroom teachers, thus allowing a different set of voices to be heard.

The goal of this series is to use teachers' accounts of classroom inquiry to make public and explicit the processes of doing research in classrooms. Teaching is a complex, multi-faceted task, and this complexity is often not captured in research articles. Our goal is to illuminate this complexity. Research that is done in classrooms by and with teachers is necessarily messy, and our stance is that the ways in which this is so should be articulated, not hidden.

Identifying Authors

Using the participants of the Albuquerque conference as a starting point, the editorial board for the series generated a list of teachers, projects, and university faculty who we knew had been engaged in classroom research themselves or might know of others who had. Through personal contacts with these individuals, and those they led us to, we compiled a list of potential authors, the nature of their work, and relevant background information. For each volume, we attempted to generate a set of potential authors and topics that would span aspects of the *Principles and Standards for School Mathematics* (NCTM, 2000), such as the content standards, process standards, and principles. We also tried to find a set of pieces that would span various roles that teachers might play in research—principal researcher, co-researcher, research participant, and consumer of research. We then invited authors for specific chapters in accordance with the categories outlined above. Authors were specifically asked to highlight the following aspects of their work in their manuscripts:

- mathematical content and processes that were addressed (using the *Principles and Standards* document as a guide);
- demographic data on school/student population/community (as it pertains to the research);
- the authors' role in the research and what the experience was like for them;
- data sources for the work (incorporated into the narrative as appropriate—for example, interview transcripts, student work, teacher reflections, summary of a class session);
- explanation of the data analysis process (How did the authors make sense of their data?);
- articles that influenced their work (rather than a full literature review, references to work that may have sparked their curiosity, helped them think of data collection methods, contrasted their findings, or was helpful in another way); and
- implications (How did this influence the authors' subsequent teaching? What might others take away from this study?).

We particularly stressed the need for the manuscripts to reflect disciplined inquiry and for claims to be based on evidence.

The Review Process

The editorial board took great care with the review process in an effort to be respectful of the varied experiences the authors had with writing for publication. Initial outlines and drafts were reviewed and commented on by the volume editor to ensure that the general flavor of the manuscript met the goals for the series. For the grades 6–8 volume, more polished drafts were reviewed by two reviewers—one teacher and one university teacher educator. To the extent possible, we solicited reviewers who had experience with teacher inquiry. We asked reviewers to be sensitive to the fact that the authors were making themselves vulnerable by sharing their research with a wider audience, and we reminded them that the type of research reflected in the manuscripts is much messier than research that is traditionally reported in journal articles. Thus, we asked the reviewers to approach their task from a mentoring frame of mind, striving to both support the authors and strengthen the manuscript. Feedback from reviewers was not given to authors verbatim; instead it was compiled by the editors to ensure that the authors received a consistent and helpful message from the review process.

HOW THIS VOLUME MIGHT BE USED

This volume, and others in the *Teachers Engaged In Research* series, was written primarily for teachers who have or who are being encouraged to develop an awareness of and commitment to teaching for understanding. The research findings presented in these chapters suggest instructional implications worthy of teachers' consideration. Authors have often described instructional practices or raised issues that have the potential to broaden views of teaching and learning mathematics. Embedded in the chapters are interesting problems and tasks used in the authors' work that teachers could use in their own classrooms. A hallmark of good research is its connection to the extant literature in the field, and the authors of this volume have themselves drawn from the research literature to inform their work. The reference list accompanying each chapter can be a useful resource and should not be overlooked. In addition, this series showcases the variety of ways teachers can become engaged in research and it is hoped that readers will recognize that teacher research can be accessible and potentially beneficial to their own work in the classroom. The volumes could be particularly useful for teacher study groups, courses for preservice and inservice teachers, and professional development activities.

This is not to imply, however, that this series is intended only for teachers. The volumes are interesting, informative resources for other researchers, policy makers, school administrators, and teacher educators as well. In particular, they provide an opportunity for those outside the classroom to gain insight into the kind of issues that matter to teachers, the ways in which those issues might be researched, and the contributions that research can make to the work of classroom teachers.

OVERVIEW OF CHAPTERS

This volume contains 11 studies of teachers engaged in research in their classrooms, and in some cases beyond their classrooms. Table 1.1 summarizes the research focus, mathematical focus, data sources, and teacher role

in the research for each of the 11 studies. In the following paragraphs, I
provide a brief overview to each chapter and I encourage you to read the
chapters to hear and understand the powerful stories of these teachers'
engagement in research.

Table 1.1.   Overview of Research in Grades 6–8 Volume

Chp #	Author(s)	Research Focus	Mathematical Focus	Data Sources	Teacher Role in Research
2	Harkness & Grant	Studied how Escher created tessellations; this research pushed the authors to think more deeply about the mathematics involved	Geometry—tessellations and transformations	Sketchbooks, and stories that tell how this experience impacted that authors personally, changed their understanding of the mathematics they used, changed their classroom practice, and changed their view of teaching and learning	Two Teacher-Researchers
3	French & Nathan	Studied (a) students' informal algebra problem-solving strategies and discourse prior to formal algebra instruction, and (b) teacher beliefs about students' intuitive solution methods to understand how beliefs affect instructional practices and how they change; Studied how teachers build on students' knowledge	Algebra—early algebraic reasoning and problems solving	Class field notes and videos; student work; student interviews; teacher interviews before/ after selected lessons; extended discussions after school (weekly); extensive sessions during intervening summers—all data above were collected each year over a 5-year period	Participating in University Researcher's Study

Table 1.1.   Overview of Research in Grades 6–8 Volume

4	Coles & Brown	Studied the possibilities of creating a classroom culture of "becoming a mathematician", and the process through which Coles learned to teach	Algebra —as a way of thinking	Students' written work; video recordings of lessons every two weeks; lesson observation notes by Brown; post-lesson write ups by Cole; semi-structured interviews with students at start and end of term	Teacher-Researcher Collaborating with University Researcher
5	Walker with D'Ambrosio & Kastberg	Studied the development of a mathematics community in a classroom, and the development of mathematical empowerment among students	Rational Numbers— part to whole concepts	Student work; field notes; observations	Teacher-Researcher Collaborating with University Researcher
6	Gutstein	Studied the process of teaching for racial and social justice; worked to understand how to create the conditions for students to develop mathematical power and have a sense of social agency	Data Analysis — probability	Student work (mathematics and writing); student journals; participant-observations; field observations; teacher journal; focus group interviews of students; open-ended surveys; informal conversations; parent interviews	Teacher-Researcher

Table 1.1.   Overview of Research in Grades 6–8 Volume

7	Renninger, Stein, Koenig & Mabbott	Studied how to support students in thinking mathematically in three teachers' classrooms through the Math Forum's Bridging Research and Practice Project	Geometry— surface area, volume, Rational Numbers— percent, proportion	Video tapes of classroom practice prior to project onset; video tapes of classroom practice in which each participant was video taped in his or her own classroom teaching the same rich, nonroutine, challenge problem; archived discussions; video tapes of workshops; draft and final copy of the videopaper on which participants worked	Teacher-Researchers Collaborating with University Researcher
8	Thomas & Santiago	Studied in what ways can a teacher-researcher collaboration make mathematics meaningful to students in an urban school	Geometry— points to polyhedra	Video tape of unit lessons, student work, field notes, teacher's journal	Teacher-Researcher Collaborating with University Researcher
9	Smith	Studied in what ways examining student work samples could enable the author to transform her teaching practice	Measurement— volume	Teacher notes in plan book, student work samples, teacher journal reflections	Teacher-Researcher

Table 1.1.   Overview of Research in Grades 6–8 Volume

10	Callard	Studied what students learn when engaged in learning mathematics through inquiry; focused on assessing student learning during an inquiry unit	Data Analysis—measures of central tendency, Measurement—linear measures	Audio and video tapes of all classes, artifacts from classroom work, all student work, all student assessments, teacher's lesson plans and reflections, researcher field notes	Teacher-Researcher
11	Shafer & Hill	Studied the integrated nature of one teacher's (Hill) instructional and assessment practices and identified dimensions of classroom interaction that supported student reasoning; investigated the relationship between the teacher's practices and student responses on constructed-response assessment items	Algebra—recursive patterns, symbolization	Vignettes constructed from reports of classroom observations, teacher journal entries, teacher interviews over a two-year period, student results from two types of study assessments—External Assessment System and Problem Solving Assessment System	Participating in University Researcher's Study
12	DeJesús-Rueff	Studied whether characters and scripts help collaborative groups develop stronger mathematical group discourse in middle school classrooms	Mathematical discourse, Algbra—rate of change, functions	Student work, journals, reflections, audio tapes and video tapes	Teacher-Researcher

Shelly Harkness and Maureen Grant are unique, as authors in this volume, in their research in that instead of students being their study participants and their classrooms their context for research, mathematics was the participant and the world was their context. These two teachers wanted to learn more about the tessellations that both of them had been teaching to their students. They wanted to understand how Escher created his tessellations and wanted to experience the inspiration he gained from the mosaics of Granada, Spain. Their chapter tells the story of how this research experience impacted them—and changed their understanding of mathematical ideas, their classroom practice, and their view of teaching and learning.

Amy French and Mitch Nathan write about their experiences as a classroom teacher and a university educator participating in a university research partnership. Their research focused on understanding how students own intuitions were used in the midst of problem-solving situations in the classroom, and how this type of understanding could be used by the teacher to facilitate the learning of formal algebraic concepts and problem-solving methods. Their chapter reveals a classroom teacher's honesty and openness about the vulnerability she felt in having researchers in her classroom and in examining closely her practice and her students' learning.

Alf Coles and Laurinda Brown share their reflections on Alf's journey in learning to teach. Their research is focused on whether it is possible to create a school algebra culture among 11 year olds, whereby students experience a need for algebra and show some understanding of its power and use. The chapter tells the story of Alf's efforts to create communities of inquiry through several years of teaching.

Vicki Walker, Beatriz D'Ambrosio and Signe Kastberg also focused on inquiry and developing a mathematical community. They share the early stages (what they call the planting phase) in this development and then discuss the established community. Their chapter tells the story about mathematical empowerment for all students in the class, and in particular for the "non-compliant, underachieving students" in the class.

Eric Gutstein writes about his research studying student learning and his own experiences of teaching mathematics for social justice. He focused on studying the process of teaching for racial and social justice—trying to understand how to create the conditions for students to develop mathematical power and have a sense of social agency.

Ann Renninger, Susan Stein, Judy Koenig and Art Mabbott reflect on their experiences participating in the Math Forum's Bridging Research and Practice Project and how this impacted their classroom practice. Their chapter tells the story of these authors' learnings as they encourage and promote students in developing mathematical thinking.

Christine Thomas and Carmelita Santiago share their classroom research which focused on ways of engaging urban learners in learning mathematics in meaningful ways through collaborative activities. They write about a 21-day geometry unit titled "From Point to Polyhedron: The Icosahedron" in which students were engaged.

Tracey Smith writes about her research in trying to transform her teaching practice. Her narrative illustrates some of the inherent complexities she faced when she decided to change her approach to the teaching and learning of mathematics. She was influenced by readings as part of her graduate study, and became aware of using open-ended tasks as a way of integrating assessment with instruction. She decided to transform her practice by collecting and analyzing data that would help her understand the ups and downs of implementing open-ended tasks. Her chapter details her action research cycles of plan, act, observe, reflect, and then re-plan.

Cindy Callard reflects on her journey as a teacher and researcher, and shares research that she undertook to document and analyze what students can learn in a mathematics classroom informed by an inquiry approach. She engaged her students in an inquiry unit and focuses on how she assessed their learning during this unit, and what she learned from this assessment.

Mary Shafer and Annette Hill discuss their research in which they examined the integrated nature of Hill's instructional and assessment practices and identified dimensions of classroom interaction that supported student reasoning. They then investigated the relationship between the teacher's practices and student responses on constructed-response assessment items.

Marcia DeJesús-Rueff shares her story of her quest to develop stronger mathematical discourse from collaborative student groups. She undertook this research to address what kinds of teaching and learning strategies would help her students work through their mathematical misconceptions. A strategy she focused on involved students developing mathematical discourse skills through role playing.

COMMON THEMES

Through the chapters in this volume we learn about the research foci and/ or questions that these classroom teachers are interested in examining, the mathematics content through which they engaged their students in these explorations, the data sources they used to make sense of their focus and questions, and their roles in the research.

Mathematical thinking, mathematical communication, mathematical power, engaging students in inquiry, and examining teacher practice are common threads through many of the chapters in this volume. One way of describing the research foci and/or questions that these authors investigated is through the categories of teachers' mathematical thinking, students' mathematical thinking, teachers' practice, and sociocultural environment. While one study (Harkness & Grant) focused primarily on deepening teachers' mathematical thinking, eight studies examined students' mathematical thinking and ways of developing that (Callard; Coles & Brown; French & Nathan; Renninger, Stein, Koenig, & Mabbott; Shafer

& Hill; Smith; Thomas & Santiago; Walker, D'Ambrosio, & Kastberg). Examining teachers' practice was a focus in nine out of 11 of these research studies (Callard; Coles & Brown; DeJesús-Rueff; French & Nathan; Harkness & Grant; Renninger, Stein, Koenig, & Mabbott; Shafer & Hill; Smith; Thomas & Santiago). Four of the studies reported in this volume have a research focus and/or question involving sociocultural environment (Coles & Brown; DeJesús-Rueff; Gutstein; Walker, D'Ambrosio, & Kastberg).

The mathematics content in which teachers and students are engaged in these research studies includes geometry (Harkness & Grant; Renninger, Stein, Koenig, & Mabbott; Thomas & Santiago), rational numbers (Renninger, Stein, Koenig, & Mabbott; Walker, D'Ambrosio, & Kastberg), algebra (Coles & Brown; DeJesús-Rueff; French & Nathan; Shafer & Hill), data analysis (Callard; Gutstein), and measurement (Callard; Smith). Mathematical discourse is the primary focus of one chapter (DeJesús-Rueff).

It is fair to say that the authors of the research studies reported in this volume "built their research methodology upon the reflective practices they already engaged in as teachers" (Dadds & Hart, 2001, p. 145). All of the researchers used methods of observation and inductive analysis to investigate their research questions and aims, and many of the researchers collected data through classroom observations, student work, and audio and/or video recordings of classroom lessons.

The chapters in this volume show teachers playing various roles in research studies. Several authors were the sole teacher-researchers in their projects (Callard; DeJesús-Rueff; Gutstein; Smith). One set of authors were teachers collaborating with each other as researchers (Harkness & Grant). A number of chapters were written by teachers and university educators who collaborated together as researchers (Coles & Brown; Renninger, Stein, Koenig, & Mabbott; Thomas & Santiago; Walker, D'Ambrosio, & Kastberg). Two chapters were written by teachers and university educators about the classroom teacher's participation in the university educator's research study (French & Nathan; Shafer & Hill).

An overarching theme through all the chapters in this volume is the learning and professional development that occurs through teacher research. What these authors learned about student learning and their own teaching practice far exceeded the focus of their particular research questions. For some, the research validated their beliefs and instructional practices; for others, it deepened or extended their understanding of mathematics, or raised their expectations of students' capabilities. For all, it is fair to say, their research increased their awareness of how students come to know and understand mathematics, and enabled them to gain insight into the complexity of teaching. Lampert (2001) noted, "One reason teaching is a complex practice is that many of the problems a teacher must address to get students to learn occur simultaneously, not one after another" (p. 2). We, as readers, gain a window into these teachers' research within and about the complexity of classroom teaching.

Collaboration is another overarching theme that threads throughout this volume. Almost all of the teacher authors of these chapters were engaged in collaborative research, usually with a university colleague. School-university collaborations provide one important context for teacher development, and for anchoring research into this development (Cochran-Smith & Lytle, 1993). Involving varieties of negotiations and compromises regarding insights, understandings, and theories, such projects allow teachers and researchers to construct models that reflect the day-to-day decision-making of teachers in complex classroom settings (Chandler-Olcott, 2001). The collaborations described in these chapters often engaged researchers in analyzing student learning and tasks. Analyzing mathematical tasks can be a powerful tool in understanding the mathematical thinking required in the task and the difficulties faced by students in achieving this thinking (Stein, Smith, Henningsen, & Silver, 2000). Through these chapters, we gain insight into these teachers' reflection on teaching and their analysis of teaching tasks, and we hear their stories of how they came to understand better how to mediate their students' learning.

AN INVITATION

For the most part, what you will read in this series are teachers' accounts of practice in which they have made public various aspects of their teaching and their beliefs about teaching and learning. You might be tempted to examine their instructional approaches with a critical eye or to question the rigor of their research methods. In fact, it is expected that published research stand up to such scrutiny. I encourage you, however, to also regard the work presented in this volume with respect (and perhaps even awe) and to champion the accomplishments of these teachers in studying and learning from their classroom practices and their students. They deserve no less for the risk they have taken in opening their research to public inspection. Learn from what they have presented. If you are a classroom teacher, let it inspire you to examine the questions of practice that interest or puzzle you. If you are a university teacher educator or researcher, let it be the impetus for you to engage in collaborative research with a classroom teacher.

NOTE

1.   This chapter was written, in part, as a result of collaborative conversations with the editors of the other volumes of this series and the series editor. I wish to acknowledge the contributions of Cynthia Langrall, Stephanie Smith, Laura Van Zoest, and Denise Mewborn.

REFERENCES

Chandler-Olcott, K. (2001). Teacher research as a self-extending system for practitioners. *Teacher Education Quarterly, 29* (1), 23–38.

Cochran-Smith, M., & Lytle, S. (1993). *Inside/outside: Teacher research and knowledge.* New York: Teachers College Press.

Dadds, M., & Hart, S. (2001). *Action research in education.* New York: Routledge/Falmer.

Grouws, D. A. (Ed.). (1991). *Handbook of research on mathematics teaching and learning.* New York: Macmillan.

Lampert, M. (2001). *Teaching problems and the problems of teaching.* New Haven, CT: Yale University Press.

Lester, F. K. (Ed.) (in preparation). *Handbook of research on mathematics teaching and learning* (2nd ed.). Greenwich, CT: Infoage/National Council of Teachers of Mathematics.

National Council of Teachers of Mathematics. (2000). *Principles and standards for school mathematics.* Reston, VA: Author.

Stein, M. K., Smith, M. S., Henningsen, M. A., & Silver, E. A. (2000). *Implementing standards-based mathematics instruction: A casebook for professional development.* New York: Teachers College Press.

CHAPTER 2

MOSAICS OF GRANADA TO THE MATHEMATICS OF ESCHER'S GEOMETRY

Making Connections

Shelly Sheats Harkness
University of Cincinnati, Cincinnati, OH

Maureen Grant
Metropolitan School District of Washington Township, Indianapolis, IN

In the same room as the design on the previous page that took about one hour and 45 minutes. Sitting in the same chair. Just finished the pattern below. The guard tried to talk to me in Spanish. I said, "No hablo Espanol..." and he replied "Bonita." "Gracias." This room is full of geometric shapes. I could probably spend all day here...the window grills, the wooden ceiling, the niches, the floor, the arabesque writing...Absolutely beautiful! (Shelly's sketchbook, July 19, 1997)

Teachers Engaged in Research
Inquiry Into Mathematics Classrooms, Grades 6–8, pages 17–36
Copyright © 2006 by Information Age Publishing

Throughout Spain, we found evidence of the Moorish influence on design. Some of the tiles, I felt, were there for the tourists, but much of it, especially in the older areas, appeared genuine. Tiles appeared in entryways to apartments and restaurants, around doorways and balconies, and along walls. Islamic designs appeared on wooden doors and in the grillwork on windows and security gates. Even the logos on some of the businesses (especially banks) seemed to show the Moorish influence. Gaudi, in Barcelona, also demonstrated the use of Moorish-inspired designs. (Maureen's sketchbook, July 24, 1997)

INTRODUCTION

Our research was not what you might consider typical classroom research. The research questions in our study, entitled "Inspirations from Spain: Mosaics of Granada to Escher's Geometry," emerged as ways to frame the curiosities that intrigued us about the geometry we were already teaching. In some ways, the "participant" we studied was the mathematics and our classroom included the world around us.   In the beginning, our study occurred outside of school, the result of Lilly Teacher Creativity Fellowship grants we were awarded during the summer of 1997. The two of us spent the summer "doing" geometry and making connections between what we knew and what we wanted to know more about.

The data we collected included the many learning experiences we participated in, the journals we kept throughout the experiences, and the conversations we had with one another about these experiences. In July we traveled to Spain to collect additional data in the form of geometric sketches. We sat on the floors of ancient historical palaces – such as the Alcazar in Seville, the Alhambra in Granada, where King Ferdinand and Queen Isabella met with Christopher Columbus prior to 1492, or the La Mezquita in Cordoba, famous for its 850 red-and-white striped double horseshoe arches that raise the height of the ceilings in this mosque or cathedral depending on the ruling culture of the time – and sketched the same tessellating ceramic tiles that inspired the artist M.C. Escher.

Throughout these experiences we were learning geometry with understanding; we were building new knowledge from experience and prior knowledge (The Learning Principle, National Council of Teachers of Mathematics [NCTM], 2000). Van De Walle (2004) defines understanding as a measure of both the quality and quantity of connections that an idea has with existing ideas. An individual's understanding exists along a "continuum of understanding," with one end a very rich set of connections and the other end a place where ideas are completely or nearly isolated (Van De Walle, 2004). Skemp (1976) refers to the rich interconnected end as *relational understanding* and the other end, rotely learned knowledge, as *instrumental understanding.* The two of us were moving from instrumental to relational understanding of the geometry we were already teaching in our middle school classrooms as we began to "recognize and build connections among mathematical ideas, understand how mathematical ideas interconnect and build on one another to produce a coherent whole, and recognize and apply mathematics in contexts outside of school" (Connections Process Standard, NCTM, 2000, p. 402) when we immersed ourselves in finding answers to our research questions.

This is the story of our Lilly Teacher Creativity Fellowship experience and its impact on our own relational understanding of the geometry we were teaching. As we immersed ourselves in *doing* geometry through connections and realized the impact on our own understanding during that summer experience, we wanted the same experiences for our students, to experience geometry by doing and making their own connections.

This chapter is written in the form of a narrative. Clandinin and Connelly (1990) describe the purposes of writing narratives for a larger audience: "Narrative inquiries are shared in ways that help readers question their own stories, raise their own questions about practice and see in the narrative accounts stories of their own stories. The intent is to foster reflection, storying, and restorying for readers" (p. 20). It is our hope that this story will help teachers envision their own stories and the research that they might engage in related to the mathematics they want to know more about.

THE RESEARCH IDEA EMERGED

During the 1996–1997 school year, we were mathematics teachers at an urban middle school in a large Midwestern city. The school population, about 900 students, included sixth, seventh, and eighth graders from a range of socio-economic and cultural backgrounds. Approximately 45% of the students described themselves as White/non-Hispanic, 45% as Black/non-Hispanic, and 10% as Other ethnicities. Forty percent of the students qualified for free or reduced lunch. The school had just created a position for an English-as-Second-Language teacher as our enrollment of students who spoke Spanish and Russian increased significantly.

Maureen was Mathematics Department Chairperson and taught eighth grade mathematics. It was only her second year at the middle school level; prior to this teaching assignment, she taught mathematics for thirteen years at the high school level. After teaching mostly eighth- and ninth-grade mathematics classes in a junior high school for nine years, Shelly was teaching sixth grade for the first time in a new school district. We both felt a bit like outsiders and so it seems that part of our initial motivation to apply for Lilly Teacher Creativity Fellowships emerged because we each needed a friend in this large middle school, someone we could talk to about all kinds of normal issues that arise in the course of a day in the lives of teachers.

One day in October, we found the Lilly Endowment Teacher Creativity Fellowship Program application in our school mailboxes. This program was "a summer renewal opportunity for Indiana public school teachers" (Lilly Endowment program announcement, 1996). As one purpose of the program, the Lilly Endowment aimed to support a creative project that would personally revitalize and intellectually expand teachers. A second purpose was to extend the benefits of teachers' personal renewals to their students. The Endowment committed to grant up to 80 awards in the amount of $4500 each and encouraged teams of teachers to submit collaborative proposals. Each teacher on the team would receive $4500. Public school classroom or resource teachers, guidance counselors, and library or media specialists who had at least three years of experience in Indiana were eligible to apply. Applicants were required to make a full-time commitment for eight weeks during the summer of 1997.

After reading through the application, Shelly approached Maureen and asked if she was interested in applying for a grant; it would be something we could do together. Shelly had noticed M. C. Escher posters on Maureen's classroom walls and she knew that Maureen's students created Escher-type tessellations using HyperStudio. Shelly's students had created Islamic Art designs as an activity in a geometry unit. We knew that Escher copied into his sketchbook many of the tile designs at the Alhambra Palace in Granada, Spain, in the year 1936, and that he credited this work as inspiration for his own tessellating artwork.

The notion to study Escher's artwork and link it to Islamic Art emerged as the result of a brainstorming session in Maureen's classroom during after-school hours. As we covered the chalkboard with a web of connected topics to study, the possibilities seemed limitless. We wanted to know more about the connections between topics such as: (a) Escher's artwork and the mathematics he used to create it, (b) Islamic Art and Islamic culture, (c) Spanish history, (d) Spanish and Islamic architecture, (e) the Moors who inhabited Spain prior to Queen Isabella and King Ferdinand, (f) Washington Irving (1851) who wrote the book, *The Alhambra*, (g) connections to transformational geometry, and others. The design of our research and the application for the Lilly Teacher Creativity Fellowship incorporated ideas from this web mapping that covered Maureen's chalkboard.

As mentioned earlier, we both taught geometry lessons by using connections to art. When Maureen taught honors Geometry at the high school level, her students explored different techniques for creating tessellations in the style of M. C. Escher. Beginning with a tessellating polygon such as a square, rectangle, triangle or hexagon, students used tracing paper to alter the sides of the basic polygon through the use of translations, rotations, or reflections to create a realistic shape that would tessellate. She slightly modified the tessellation project for her middle school students as they utilized a cut-and-paste technique to alter the sides of squares and hexagons to create designs with translations or rotations. During Maureen's second year of teaching at the middle school, she taught her students how to use a computer software package, HyperStudio, to create tessellating designs by altering the sides of a square with translations or rotations. During the time that Maureen was completing her master's degree work in mathematics, she wrote a paper for an abstract algebra class entitled, "The Use of Wallpaper Groups in the Work of M. C. Escher." Thus, she was able to supplement the teaching of the techniques for creating tessellations with a discussion of the life and artwork of M. C. Escher.

Shelly had been refining her geometry and Islamic Art unit for about five years. She used a packet of materials from the Metropolitan Museum of Art, *The Mathematics of Islamic Art* (Wasserman, 1979), which included slides of Islamic artifacts from the museum and some notes for "teachers of mathematics, social studies, and art." She started by telling her students about Islamic history and culture, emphasizing the notion that Muhammad preached against idolatry and early religious leaders of Islam forbade artists to create drawings of humans or animals because only Allah created life. Throughout history some artists did not adhere to this interpretation but a strongly nonfigurative art form developed. Instead of using art to illustrate religious subjects, Muslims became the great patternmakers of art history (Wasserman, 1979). Objects and buildings were covered with complex patterns. Three kinds of patterns were prevalent in Islamic Art: (a) vegetal designs, (b) calligraphic decoration that conveyed the word of Allah, and (c) repeating geometric shapes as "unity in multiplicity," an Islamic religious principle. As the culminating activity, students in Shelly's classes created designs with straightedges and compasses and based on the five characteristics of Islamic Art:

- They are made up of a small number of repeated geometric elements.
- They are two-dimensional. (Figures seem as though they are flattened.)
- They radiate symmetrically from a central point.
- They are not designed to fit within a frame.
- They are constructed from patterns of circles. (Wasserman, 1979)

The unit seemed to engage some reluctant learners, especially students who liked art and had not seen connections to geometry and art before.

Some students who had not experienced much success in mathematics throughout the school year created very intricate and mathematically rich designs. Shelly focused mostly on helping students use and understand the language of geometric terms such as polygons (including squares, rectangles, triangles, pentagons, hexagons), parallel, perpendicular, and tessellations. For the most part, she was pleased with the designs and the unit as a whole although she still felt there were other mathematical big ideas that students could have explored.

THE APPLICATION

As the application due date—January 3, 1997—approached, we decided to meet during our winter break, before the beginning of second semester. With lots of ideas and a few reference books, we sat in front of the computer and outlined what we wanted to accomplish during our proposed summer research experience. We include a section of the description of our proposal so that readers will understand how it framed our thinking:

Islamic Art is rich in geometric form. The Greek Pythagoreans, who felt that numbers ruled the universe, were an important influence on Islamic thought. The Islamic artists, therefore, were trained in geometry and their philosophy saw mathematics as essential to understanding the universe. The Alhambra, a thirteenth century palace in Granada, Spain, is one of the finest examples of the precise mathematical art of Islam that exists today. In the fall of 1922, the Dutch artist Maurits Cornelis Escher traveled to the Alhambra and became fascinated by the beautiful mosaic tiles. He returned in the spring of 1936 and spent several days making sketches of the tessellating patterns—an arrangement of figures that fills a plane without gaps or overlapping—on the floors, walls, and ceilings. Although Escher made a few attempts at creating tessellations in his early work as a graphic artist, his urge for "filling the plane" became a major preoccupation after his second trip to the Alhambra. Escher once stated, "...I often seem to have more in common with mathematicians than with my fellow artists" (Escher, 1960, p. 9). Behind Escher's intriguing patterns is a rich mathematical background that includes topics such as symmetry, group theory, and tessellations.

The Curriculum and Evaluation Standards (1989) of the National Council of Teachers of Mathematics require that, "Students should have numerous and varied experiences related to the cultural, historical, and scientific evolution of mathematics so that they can appreciate the role of mathematics in the development of our contemporary society" (p. 5). As educators, we have used examples of Islamic Art and the work of M. C. Escher to enhance the instruction of geometry. The purpose of this project is to more fully immerse ourselves in a multicultural, interdisciplinary study of these topics...As we have done the initial research necessary to prepare this proposal, we have been amazed and excited by the amount of mathematics that we have uncovered. It is our feeling that our excitement over this project will also inspire our students. Specifically, this project will help us to improve instruction that will help our students:

- understand and apply geometric properties and relationships;
- connect mathematics to other subjects and to the world outside the classroom;
- consider the mutual respect, pride, and understanding that come from the knowledge that all cultures have contributed to mathematics

We plan to thoroughly research the history, art, architecture, and mathematics of Moorish Spain. We will study the mathematics involved in Escher's tessellations and research Escher's drawings of the mosaic designs found in the Alhambra and La Mezquita Mosque in Cordoba, Spain. While visiting these locations, we will photograph and sketch the same designs that inspired Escher...

It is our goal to use the information gathered during our initial research and our trip to Spain to develop a comprehensive interdisciplinary, multimedia lesson about the Moorish influence on mathematics and the art of M.C. Escher that we can share with our students, their parents, and the Lilly Endowment. It is our goal to help all students gain an appreciation for the beauty of mathematics and a realization that mathematics is everywhere. (Excerpt from our application for the Lilly Fellowship Program)

Our research questions took the form of specific topics we wanted to know more about and also as general questions that framed our entire study. These general questions included the following:

- What would we learn about the mathematics of Escher's artwork by *doing* – sketching the same tiles that inspired his tessellations?
- How would what we learned – our subject matter knowledge – impact our teaching and, ultimately, the experiences we created for our students?
- What new connections, both big ideas of mathematics and cultural and historical connections, to Islamic Art and Escher would we find as we were immersed in this research?

THE LETTER ARRIVED

March 12, 1997

Dear Ms. Harkness [and Ms. Grant]:

It gives me great pleasure to inform you that the panel of out-of-state judges for Lilly Endowment's Teacher Creativity Fellowship Program has selected you as fellows for the summer of 1997...On behalf of the Lilly Endowment Board of Directors, may I extend hearty congratulations to you on receiving this fellowship and best wishes for a productive and rewarding summer.

Needless to say we had huge grins on our faces at school the next day. We knew that the teaching challenges we faced during the next few months would be manageable because of our excitement about the summer work ahead. In June, the school year ended and the buses full of students pulled out of the parking lot to the tune of the *Hallelujah Chorus* blasting over the public address system, an annual tradition at our middle school. We could barely contain our enthusiasm as we anticipated the excitement of spending time—such a valuable commodity in the life of a teacher—immersed in mathematics and a culture different from our own.

THE SUMMER EXPERIENCE

Prior to our trip to Spain, we studied the mathematics involved in Escher's tessellations and researched Escher's drawings of the mosaic designs found at the Alcazar, the Alhambra and La Mezquita. Shelly interned at an architectural firm for two weeks. Together, we worked alongside high school students who created mosaics that are now displayed along a walking and biking trail. We found a book about patchwork quilts made with Escher designs and studied the patterns. A four-week course in conversational

Spanish helped us learn key words and phrases we could use when we traveled. One highlight of our study included a guided tour of an Islamic center near our hometown. We bought books such as *Escher on Escher* (Escher, 1986), *Islamic Designs for Artists and Craftspeople* (Wilson, 1988), and *M. C. Escher Visions of Symmetry* (Schattschneider, 1990), to name a few. We also purchased graph paper sketchbooks, colored pencils, and lots of film for our cameras.

The day of our anticipated journey arrived. We left for Madrid in the middle of July and spent two days exploring the city. Highlights of our sightseeing in Madrid included: (a) Picasso's *Guernica*, (b) the Prado Museum's masterpieces by Zurbaran, El Greco, Velaquez, Rubens, and Goya, (c) lunch at Sobrino de Boutin, an Ernest Hemmingway "haunt" which he mentioned in two of his novels, and (d) a visit to Iglesia de San Nicolas, a church with distinctive Moorish brickwork and horseshoe arches.

Leaving Madrid, we traveled south through the landscapes of Don Quixote's *La Mancha* to Cordoba. We followed Escher's footsteps by touring the La Mezquita. In complete awe of the fact that we were actually there, we photographed and videotaped this crowning Muslim architectural achievement, famous for its 850 red-and-white striped double horseshoe pillars that Escher sketched in shades of black, gray, and white.

From Cordoba, our next stop was Seville. A climb to the top of the Giralda Tower, originally built as the minaret for Seville's mosque and the only part not destroyed when the cathedral was built in the 1400's, afforded us a spectacular view of the city. This cathedral claims to contain the remains of Christopher Columbus with his tomb mounted on four statues. We spent an afternoon sketching geometric designs found in the mosaic tiles of the Alcazar, a 14th century Mudejar ("Christianized" Moorish-influenced architecture) palace. During our last evening in Seville, we enjoyed the sights and sounds of flamenco, a dance influenced by Jewish, Islamic, and gypsy artists.

From Seville, we passed through the smaller cities of Jerez and Ronda on our way to Torremolinos. We took a day trip from there to Gibraltar, a British colony. Gibraltar's location at the narrow entrance from the Atlantic Ocean to the Mediterranean Sea led to its seizure by the Moors in 711, prior to their conquest of Spain.

Away from the Mediterranean Sea and into the mountains, we arrived at our main destination, Granada. On our first day in Granada, a tour guide took us through the Alhambra explaining the history of this Moorish citadel. In one of the rooms of the Alhambra, Christopher Columbus asked Queen Isabella and King Ferdinand for money to find the "New World" and we were standing there!

The next two days, in our own paradise, we sat and sketched the geometric designs that, to us, had previously been only pictures in a book, *M. C. Escher Visions of Symmetry* (Schattschneider, 1990), but were now reality. Prior to our visit to the Alhambra, when we looked at pictures of the tile designs, we wondered if each of the colored shapes in the tile designs were separate tiles or if the designs were stamped onto larger tiles that were then pieced together. Our opportunity to look at the tiles closely revealed that the designs were indeed created with separate tiles. We were amazed at the precision of the craftsmanship and the amount of mathematical knowledge that was necessary for the creation of the tile designs. We made a mental note to pass these lessons along to our students.

The palace was even more beautiful than we anticipated. For us it was a dream come true! And, even more startling was the fact that these designs were much harder to copy than we anticipated. We abandoned our rulers for credit cards. What we mean by that is that the rulers became cumbersome and the sides of our credit cards became the perfect tool to draw the edges of the polygons we sketched in our books. After the first day, we realized that there were too many designs to capture in our books, given the amount of time it took to copy each one to scale and then color them, so we went to different palace rooms and started copying designs and just noting the colors of the individual polygons. In our hotel room at night we sat

Figure 2.1.   Wall tiles in the Sala de la Barca in the
            Alhambra in Granada, Spain.

Figure 2.2.   Sketch of the wall tiles in the Sala de la Barca
in the Alhambra in Granada, Spain.

coloring designs. If people had seen us, they would have thought us completely nuts!

Because we knew that time was running out and we could later use the developed prints to create more designs in our sketchbooks, Maureen photographed the designs, especially the patterns that looked like what we described as intertwining Celtic knot work. We left the Alhambra with a deeper appreciation for Escher's attempts to sketch the same tiles that we sketched and then transform his notions of symmetry into tessellations that included curves instead of merely straight-edged polygon shapes and included figures such as reptiles, birds, and fish rather than the strictly abstract geometric designs found in traditional Islamic art.

The final city of our travel was Barcelona. There, we became enthusiasts of Antoni Gaudi, master architect. We had not even considered Gaudi's work in our original proposal to Lilly. It was through our study before the trip began that we learned of his work and decided to visit some of his famous architectural achievements.   The two most notable Gaudi landmarks we visited were the unfinished cathedral, *La Sagrada Familia*, and the Art Noveau extravaganza, *Parc Guell*. Moorish influence was evident in many of the architectural details and mosaics found in Gaudi's designs and at the *La Sagrada Familia* a mathematical magic square, carved on one of the massive stone reliefs at the entrance to the cathedral, intrigued us.

Our studies and travels throughout the summer of 1997 created opportunities for us to learn by doing mathematics. When we sketched the tiles
that inspired Escher we were doing mathematics and using mathematical
language to describe our sketches. The Geometry Standard for Grades
6–8 (NCTM, 2000) contains expectations or objectives for all students.
This list of expectations for students includes some of the same objectives
that we had anticipated for our own learning prior to and during our trip
to Spain such as: (a) describe sizes, positions, and orientations of shapes
under informal transformations such as flips, turns, slides, and scaling, (b)
examine the congruence, similarity, and line or rotational symmetry of
objects using transformations, and (c) draw geometric objects with speci-
fied properties, such as side lengths or angle measures. When we had the
opportunity to explore, play, and make conjectures about the mathematics
we were using our own knowledge grew by leaps and bounds. We
exceeded our own expectations as we immersed ourselves in the mathematics. The data we collected throughout the summer—learning experiences we participated in, journals we kept, sketches of the tiles, and
conversations we had with one another about these experiences—helped
us answer our research questions.

RESULTS

Mathematics We Learned

Ma (1999) suggests that to improve mathematics education for students in
the United States, teachers' knowledge of mathematics must first improve.
The Teaching Principle (NCTM, 2000) includes the notion that effective
teaching requires teachers to continually seek improvement. Reflection
and analysis are often individual activities but by teaming with experienced
and respected colleagues, teachers can enhance these aspects of their practice. Our own knowledge of the geometry we were already teaching
improved throughout our summer experiences as we collaboratively
worked together to answer research questions that were of interest to both
of us. We include some of the mathematical notions that we tinkered with
and some of the conclusions that we made in this section.

As a result of sitting and sketching we came to the realization that the
square-grid paper in our sketchbooks worked fine for some designs but not
others. Why was it harder to copy some designs and not others? Finally, we
realized that the designs were based on grids that were both square and triangular. For designs with eight-point stars and octagons the paper we had
worked fine but for designs with six-point stars and hexagons the square
grid paper made copying these designs nearly impossible. Aha! Although

we only took square grid paper with us to Spain, we realized that, ideally, we needed both square and triangular grid paper. That was when we started classifying designs as based on triangles or based on squares. And later, after we returned home and looked closely at Escher's designs we began to classify his tessellations as to whether he used square or triangular grids as the underlying structure.

We also began to think more about shapes that tessellate and why they tessellate. The importance of the 360-degree rotation around a point became much clearer to us. Through exploration we realized why only some regular polygons tessellate and why some combinations of regular polygons tessellate. For example, regular octagons will not tessellate by themselves but they will tessellate when combined with squares. When the vertices of two regular octagons (each with angle measures of 135 degrees) and the vertex of a square meet, the sum of the three angles is 360 degrees; you can fill a paper with regular octagons and squares with no gaps or spaces in between. We experimented with many combinations of regular and nonregular polygons to understand this relationship. This must have been a notion that Islamic artisans understood. As we later learned, it was an idea that Escher also grappled with.

When we first looked at the pictures of the tiles in books, we anticipated that they would be fairly easy to copy. However, this was not true for many designs. It seems, at least in the beginning, that we looked only at small segments of the overall designs and copied one polygon at a time onto our graph paper. When this became time consuming, tedious, and confusing we began to view the entire design with a *symmetry lens*. We looked at how the designs translated, rotated, and reflected and let the symmetries guide our placements of the polygons within the designs. In other words, we looked for symmetry patterns and began copying these patterns instead of just individual polygon tile pieces within the patterns. This was a much more efficient method of transferring the tile designs to our sketchbooks. As we thought of this later, it was as if we made mental maps to organize our activity. We shifted from sketching models *of* thinking to sketching models *for* thinking (Gravemeijer, 2000). Gravemeijer (2000) says that, "The shift from *model of* to *model for* concurs with a shift in the students' thinking, from thinking about the modeled context situation, to a focus on mathematical relations" (p. 9). It was the dissonance and struggle we felt with the first method that led to the symmetry lens method.

Impact on Our Practice and Student Learning

Ball (1991) and other researchers have conducted research related to teachers' knowledge of mathematics and their classroom practice. Teachers' knowledge of mathematics enables them to structure classroom experiences – make judgments about which student suggestions to pursue, develop tasks that encourage certain kinds of exploration, and conduct fruitful class discussions – that engage and empower students (Ball, 1991). As a result of our own experiences we came to realize the potential of NCTM's (2000) Curriculum Principle, "A curriculum is more than a collection of activities: it must be coherent, focused on important mathematics, and well articulated across the grades" (p. 14). Based on our own new connections and the big ideas we learned, we made additions to our geometry units that helped our students make better connections.

Maureen used a new activity with large Escher prints. She gave groups of students two or three prints, blank overhead transparencies, and markers. Each group was instructed to explore the translations, rotations, reflections, and glide reflections present in the patterns and to form a conjecture about the type of grid that they thought Escher used – square or triangular. The students' presentations as they reported back to the whole group showed evidence that they understood these terms as they defended their positions and engaged in student-to-student discourse.

A completely new addition to Shelly's geometry unit engaged the students in actually doing what we had done when we visited the Alhambra Palace. She enlarged a photograph of one of the walls of the palace and then made an overhead transparency of this photograph. Students used graph paper, rulers, and colored pencils to sketch the tessellations in the photograph as though they were sitting on the floor of the Alhambra Palace. After they finished, students showed their designs during large group discussion. Interestingly, students had different interpretations of the design because they used different scale factors, a mathematical big idea connection. For example, if the design was composed of squares and octagons some students drew their designs using one square of the grid paper to denote one square tile. Other students used four squares of the grid paper to represent one square tile in their design. We talked about the idea of scale factor and tried to determine if the different scale factors were consistent throughout the two different interpretations of the designs. Students discussed how they started and the pitfalls that they encountered as they drew the designs. The discussion was, indeed, very mathematically rich. Although Shelly has not done this, another idea she has for this activity is to show students a design based on triangles and hexagons and give them square grid paper. Will they feel the frustrations that the two of us felt under the same constraints as we were sketching the tiles at the Alhambra, and thus come to understand that there are different mathematical structures underlying the symmetry patterns in the tile designs?

When it came time to create the grids that they used to make their designs based on the five characteristics of Islamic Art (mentioned earlier), instead of demonstrating, Shelly allowed her students to have more time to experiment with the grids. With more freedom to explore they seemed less frustrated and more willing to be creative when they made their designs. She asked questions about square and triangular grids and made more connections to reflection, rotation, translation, and glide reflection. She also created lessons that allowed students to explore the question "What shapes tessellate?" and to make conjectures about "Why?" certain shapes always tessellate.

New Connections

When the 1998 NCTM Annual Convention took place in Washington, D. C., the two of us had the opportunity to visit the National Museum to see the special exhibit of Escher work, our first opportunity to see his work in person. Since our summer research, we have made presentations for the Indiana Council of Teachers of Mathematics, the Lilly Extending Creativity summer workshop, the International Baccalaureate program at a local high school, and a summer workshop for teachers in our school district (we had both mathematics and art teachers who attended this workshop!).

For Maureen, becoming a Lilly Fellow led to another opportunity two years later when she was once again teaching at the high school level and was selected as a participant in the Toyota International Teacher Program. In her proposal for the two-week study trip to Japan, an extension of the Lilly Fellowship, she wanted to study the use of geometry in the traditional art and architectural decoration of Japan. Thirty rolls of exposed film attest to the fact that everywhere she turned she found geometric designs. One specific area of interest was to study the geometric designs of Japanese family crests. She found examples of family and religious crests in museums, in a private home, and at many shrines and temples. The study of two-dimensional patterns (the so-called "wallpaper groups" or, in Japanese, "monyo gun" which means "patterns-on-traditional-clothing groups") was another focus in Maureen's search for geometric designs. Examples of two-dimensional patterns were found on traditional Japanese fabrics and also in architectural decoration. Students in a home economics course at a junior high school in Toyota City stitched a traditional two-dimensional pattern that Maureen also saw painted on a building at the Kiyomizu Temple in Kyoto. Even the walls at the entrance to an elementary school in Tokyo held an example of mathematical art, a tessellating fish design. M. C. Escher would have been proud.

As a result of her study of Japanese family crests, Maureen incorporated a new activity into her honors Geometry course which asked students to explore the characteristics of these crests. For this activity, small groups of students were given a sheet of paper that contained approximately fifty examples of a variety of Japanese crest designs. Students were first asked to classify the crests in any way that they desired and to share with the class their method of classification. The discussion that emerged from this activity included observations that were both non-mathematical (such as the fact that designs from nature such as birds, flowers, and insects are common themes in Japanese family crests) and mathematical (for instance, Japanese crest designs are usually enclosed within a circle and may have rotational or reflectional symmetry). After their initial exploration of the crest designs, students were asked explicitly to create a Venn diagram on a large piece of butcher paper and to organize the crests into four sets: designs containing (a) rotational symmetry only, (b) reflectional symmetry only, (c) both reflectional and rotational symmetry, and (d) neither reflectional nor rotational symmetry. Students realized that in classifying these real-world pieces of artwork they were required to make decisions about what constituted a significant detail in the design and what constituted a minor flaw. As they classified the designs, students began to make conjectures such as, "If a design has only one line of symmetry, then it does not have rotational symmetry" or "If the number of lines of symmetry is greater than or equal to two then the number of lines of symmetry is equal to the number of rotations of the design that are possible."

Geometric designs were only one aspect of Maureen's quest for examples of the use of mathematics in Japan. At the Edo Tokyo museum, she was pleased to find several examples of block prints that illustrated the Japanese artists' sense of the fractal aspects of nature. Her search for *sangaku* – wooden tablets engraved with geometry problems hung under the roofs of shrines and temples – ended in success at the Kitano Tenmangu Shrine in Kyoto.

Maureen's experiences in Spain and Japan led her to the realization that she could probably select any country or any culture and find examples of the use of geometry in their traditional artwork. She began to realize that many of the geometric designs also had religious connections. The first time that she walked into the church at St. Meinrad Archabbey in southern Indiana she immediately recognized the geometric design in the beautiful Italian marble floor. In response to her excited letter, a friend who is a retired mathematics professor responded, "You go to a chapel and see Sierpinski triangles?!" After many subsequent trips to St. Meinrad, Maureen wrote:

It is interesting the feeling that comes over me when I stand and look at the floor at St. Meinrad. When we [Shelly and I] were in Spain, we were studying Islamic art, which definitely has religious connections, and in Japan I was studying family crests and sangaku, which also have religious connections. To find geometric art that is connected to my own religious tradition is just amazing to me. And to stand there and look at that floor and have two seemingly disparate portions of my life (the mathematical and the spiritual) come together in one place is an awesome experience.

An interesting discovery in a book on fractals caused all of this study to come full circle when Maureen found a photograph of Escher's sketch of the cathedral in Ravello, Italy, and realized that it contained Sierpinski triangles! The web of connections continued to grow!

Maureen's broadening perspective on the use of geometry in the traditional art and architectural decoration of various cultures provided her with a new awareness and she began seeing geometric patterns everywhere—in the designs in the African mud-cloth jackets made by a local craftswoman, in the pottery and basketry at the pueblos in New Mexico that she visited one spring break, in the mosaic tile designs throughout the Basilica at the Catholic University of America in Washington D. C., on the tombstones and crypts at a local cemetery, in the architectural decoration of the buildings in the downtown area of her own city. When she shared her enthusiasm for the use of geometry in art and architectural decoration with parents at her annual back-to-school night, one parent approached her afterwards to describe a project that his architectural firm had just completed—a church within an hour of our city that incorporated the use of the Golden Ratio, Golden Spirals, and the Fibonacci sequence into its design. A second parent later sent Maureen copies of pictures that he had taken many years ago of Buckminster Fuller, the inventor of the geodesic dome among other things, when Fuller had given a lecture at his college. All of these experiences led Maureen to a growing level of confidence in incorporating the artwork and architectural decoration of various cultures into her geometry lessons and also led to a major project in which students were asked to choose a topic of their own (such as nature, toys, quilting, car hubcaps, local building or parks) and collect (preferably with photographs) examples of the use of symmetry within that topic. The posters that students created to display their examples were accompanied by a paper in which they discussed the various types of symmetry present in their photographs and pictures. Through this project, students were making their own connections between the relatively sterile geometry presented by their textbook and the beautiful examples of the use of geometry present in the world all around them.

This research led Shelly to an interest in and readings about the area of ethnomathematics. Ubiratan D'Ambrosio, the "Father of Ethnomathematics," says that ethnomathematics invites us to look into how knowledge is built throughout history in different cultural environments. Ethnomathematics is the study of modes, techniques, arts, and styles of explaining, understanding, learning about and coping with different natural and cultural environments (D'Ambrosio, 1985).   This field of mathematics includes both academic (Eurocentric) and out-of-school mathematical practices invented or used by cultural groups to make sense of or solve problems. Ethnomathematics challenges the Eurocentric view of mathematics; mathematics has been and continues to be invented in other areas of the world besides Europe. Mathematics is sometimes viewed as a universal language and yet different cultures have used different forms of mathematics throughout history. Ethnomathematics also questions the political nature of mathematics. Shelly began to consider the question, what counts as mathematical knowledge? Harris (1997) describes a cultural activity—knitting the heel of a sock. She asks why knitting the heel is not considered mathematical but the industrial problem of lagging a right-angled cylindrical pipe is considered mathematical (Harris, 1997). Shelly looked for ways to introduce more notions of beliefs about mathematics and culture into her curriculum.

After reading the book *African Fractals* by Eglash (1999), she became very interested in fractals. Eglash found fractal designs in African hairstyles, settlement designs, and artifacts. Shelly focused on fractal sessions at the last NCTM Annual Meeting, trying to attend any and all she could find in the program brochure. She bought bunches of fern leaves and introduced her students to fractals through asking questions about the leaves that helped them see the self-similarity embedded in the leaf structure. From there students copied fractal designs and then discussed the ideas of perimeter, area, and scale. Shelly became curious about connections to fractals and how she could use fractal geometry to introduce other big ideas of mathematics.

CONCLUSION

Our passion for geometry turned into a relational learning experience. On the "continuum of understanding" described by Van De Walle (2004), we feel that we moved closer to the relational understanding end. We saw many connections to other big ideas and used those connections to create opportunities for our students to make their own connections. The learning was not measurable in terms of numbers but on a rating scale we would say that it earned the highest possible score. We learned even more than we expected about the geometry we were already teaching. The impact of this experience made us think more deeply about how to change our practice to help students make their own connections to the geometry that we were teaching. Additionally, we also made new connections to big ideas of mathematics, culture, and history. Ultimately, these new connections have led us to different mathematical notions that we want to explore. The poet Rilke (1984) says,

>...to have patience with everything that is unresolved in your heart and try to love *the questions themselves* as if they were locked rooms or books written in a very foreign language. Don't search for the answers, which could not be given to you now, because you would not be able to live them. And the point is, to live everything. *Live* the questions now. Perhaps then, someday far in the future, you will gradually, without even noticing it, live your way into the answer. (pp. 34-35)

As we lived them, we found answers to our research questions and in the process of making connections our old questions were replaced by new questions and new topics we want to explore. For example, because our initial research into the tile designs at the Alhambra focused mainly on the tessellating designs, we were surprised by the amount of tile work that included knot work patterns, something we had always associated with Celtic artwork. We are now very curious about the mathematics of Celtic knot work and we look forward to finding time, a precious commodity for teachers, to collaborate, explore, and experiment with Celtic designs. More questions emerged: Are there other cultures that use knot work patterns in their traditional artwork? Had there, at some point in time, been contact between cultures that led both Moors and Celts to create similar designs or is there something innate in the human brain – some basic inclination toward order and pattern – that led the cultures to independently create these patterns?

It is our hope that through our story, other teachers can envision their own stories and the research that they might engage in related to the mathematics they want to know more about. We also look forward to envisioning more stories about our new questions and curiosities.

REFERENCES

Ball, D. L. (1991). Research on teaching mathematics: Making subject-matter knowledge part of the equation. In J. Brophy (Ed.), *Advances in research on teaching* (Vol. 2, pp. 1-48). Greenwich, Conn.: JAI Press.

Clandinin, D. J., & Connelly, F. M. (1990). Narrative and story in practice and research. In D. A. Schon (Ed.), *The Reflective Turn: Case Studies of Reflective Practice* (pp. 258-282). New York: Teachers College Press.

D'Ambrosio, U.(1985). *Socio-cultural bases for mathematics education.* Unicamp.

Eglash, R. (1999) *African fractals: Modern computing and indigenous design.* Piscataway, NJ: Rutgers University Press.

Escher, M. C. (1960). *The graphic work of M. C. Escher.* New York: Meredith Press.

Escher, Maurits C. (1986). *Escher on Escher.* New York: Harry N. Abrams, Inc.

Gravemeijer, K. H. E. (2000 August). A local instruction theory on measuring and flexible arithmetic. Paper presented at the International Conference of Mathematics Education, Tokyo, Japan.

Harris, M. (1997). An example of traditional women's work as a mathematics resource. In A. P. Powell & M. Frankenstein (Eds.), *Ethnomathematics: Challenging eurocentrism in mathematics education* (pp. 215-222). Albany, NY: State University of New York Press.

Irving, W. (1851). *The Alhambra.* New York: Thomas D. Crowell Co.

Ma, L. (1999). *Knowing and teaching elementary mathematics.* Mahwah, NJ: Lawrence Erlbaum Associates.

National Council of Teachers of Mathematics. (1989). *Curriculum and evaluation standards for school mathematics.* Reston, VA: Author.

National Council of Teacher of Mathematics. (2000). *Principles and standards for school mathematics.* Reston, VA: Author.

Rilke, R. M. (1984). *Letters to a young poet #4.* (S. Mitchell, Trans.). New York: Random House. (Original work published 1903)

Schattschneider, D. (1990). *M. C. Escher visions of symmetry.* New York: W.H. Freeman and Company.

Skemp, R. R. (1976). Relational understanding and instrumental understanding. *Mathematics Teaching, 77,* 20-26.

Van De Walle, J. A. (2004). *Elementary and middle school mathematics teaching developmentally.* Boston, MA.: Allyn and Bacon.

Wasserman, R. (1979). *The mathematics of Islamic art: A packet for teachers of mathematics, social studies, and art.* New York: Metropolitan Museum of Art.

Wilson, E. (1988). *Islamic designs for artists and craftspeople.* New York: Dover Publications, Inc.

CHAPTER 3

UNDER THE MICROSCOPE OF RESEARCH AND INTO THE CLASSROOM

Reflections on Early Algebra Learning and Instruction

Amy French
Angevine Middle School, Layfayette, CO

Mitchell J. Nathan
University of Wisconsin , Madison, WI

In the spring of 1995, when Mitchell Nathan first approached me about being part of an early algebra study, I was intrigued. I had recently spent several months with Boulder Valley School District's algebra study group—educators dedicated to fostering the growing mathematics reform movement by providing elementary, middle and high school students with meaningful access to the study of algebra. Mitch's proposal seemed like a natural next step in my teaching goals. I pondered over my final decision, however, because I had concerns about the time commitment. I

Teachers Engaged in Research
Inquiry Into Mathematics Classrooms, Grades 6–8, pages 35–54
Copyright © 2006 by Information Age Publishing
All rights of reproduction in any form reserved.

was consumed by my work—why would I want to take on anything extra? I already knew I was a good teacher, highly respected in the community for my dedication to students and effective teaching style. My evaluations were outstanding, and I had a file full of letters from grateful parents. A little, nagging fear began to emerge that maybe I would not pass muster under the university microscope. However, I ignored the disquietude and said yes. Why? Despite some minor mathematical insecurities, I felt confident enough about my practice to tackle a challenge. It had been five years since I had made any significant changes in my teaching, and I understood the value of change for the sake of growth. I saw myself teaching for only another ten years or so, and I wanted to make something of those last years. I was just beginning to see the value of the math reform movement, and needed some fresh strategies. Research sounded important and meaningful, and Mitch's initial approach to me was open, honest and collaborative.

Mitch had certain stated goals that guided the project. He and his colleagues sought to understand how students used their own intuitions during problem-solving situations in the classroom and to determine how this type of understanding could be used by teachers to facilitate the learning of formal algebraic concepts and problem-solving methods. By addressing both students' informal conceptions of algebra and teachers' beliefs about students, they hoped to develop practical ways to improve algebra instruction in middle school level classrooms. These goals shaped three specific research objectives:

1. To document students' informal algebra problem-solving strategies and discourse before students have received formal algebraic instruction;

2. To document teachers' beliefs about students' intuitive solution methods, understand how these beliefs affect teachers' instructional practices, and observe how teachers' beliefs and practices change as they learn more about the effectiveness of students' informal solution methods; and

3. To understand how knowledge of students' informal solution methods can leverage the use of more formal and more general methods for solving algebra level problems, and observe how teachers draw on new understandings of students to foster algebraic reasoning and discourse in the classroom.

Several forms of data were collected over the years. The research team examined students' written work on algebra-level problems carefully designed to allow comparisons between problems in different formats (such as matched equations and story problems). Student solutions were then analyzed for their strategies and representations and the nature of their errors. (These analyses were later published. Relevant papers for the interested reader are Nathan and Koedinger 2000a, 2000b; Koedinger and Nathan, 2004.) In addition, the team collected video tapes of classroom interactions and student discourse in small and large groups, as well as students' presentations and justifications of their solution methods. (Relevant papers include Nathan and Knuth, 2003; Nathan et al., 2002.) The research team also conducted frequent interviews with the teacher to help uncover her own rationale for classroom practices (A relevant paper is Nathan and Knuth, 2003). Finally, to obtain a broader perspective, the team administered surveys to elementary, middle and high school teachers designed to elicit their beliefs of students' mathematical abilities and development. (A relevant paper for this is Nathan and Koedinger, 2000c.)

UNDER THE MICROSCOPE

We began meeting in June of 1995, in Mitch's small office at the University of Colorado in Boulder. Crammed around a small table, I was mentally poked and probed by Mitch and two assistants, Rebekah Elliott and Eric Knuth, both doctoral students in mathematics education. From the very beginning, I never minded being questioned by Mitch. I always felt safe with him. Even though he had a mathematics degree, I felt that his main interest lay in the research. But explaining myself to the graduate students was nerve-wracking. I soon realized that being concerned about time had been only part of the reason for my hesitation to agree to this project. Those little nagging fears about my abilities grew. My minor mathematical insecurities became major. I was afraid to be found out that I lacked mathematical confidence and knowledge. I had struggled with mathematics all of my life, did not have a mathematics degree, and only by chance had become the lead mathematics teacher on a sixth-grade teaching team. While it was no secret to the school district, my principal and my colleagues that my qualifications did not include a mathematics degree (my certification is in elementary education), I was sure it would become a problem. How could I possible talk intelligently about mathematical concepts? My last math class had been a Marilyn Burns workshop in 1989 and prior to that basic college algebra in 1969!

That first summer, we talked (or rather I talked) about myself and my beliefs about teaching and learning. I rather enjoyed this at first. I explained why I had become a teacher, what made a good teacher, what was hard about teaching, and so on. I proudly described my teaching history, the horror stories of the first year, all the subjects I had taught (by this time I had experience with every discipline). My confidence and pride lessened somewhat when it began to feel like a therapy session. I was probed (using the method described by Fenstermacher and Richardson in Richardson, 1994) to explain why I felt and acted the way I did. What were my core beliefs? Why did I believe them? It seemed like every response of mine brought on another onslaught of "why." I would like to think that I was totally honest about my beliefs, hopes and fears during that first round of questioning, but I know that I felt compelled to withhold the big one—my perceived lack of mathematical knowledge. I did admit to feeling vulnerable and insecure, and revealed my own fears about mathematics as a child and young adult. I remember I even explained that one of the reasons I wanted to teach mathematics for understanding was because I did not have that experience in my own schooling, and I felt it would have made a difference in my attitudes and success with mathematics. Despite these confessions, it took time before I would admit to the level of my mathematically inadequate feelings.

In summary, I enjoyed the questioning that probed my memories and my attitudes about my early years and teaching in general. I liked the self-reflection and new insights it brought because I was confident about my instructional style and understanding of sound pedagogy. However, when the questions became more focused on the mathematics, I felt the same kind of nervousness and angst as I did when I was a child, trying to unravel a word problem and panicking that I would be unable to find the answer. That would mean I was "dumb." In that first summer of research, I did not want to appear dumb.

When school started, there was new pressure—the video camera. Now, I did not mind being filmed at first. I am a performer and I love an audience, so I was not particularly self-conscious during the taping. But I hated watching myself. I was so afraid that I would see myself saying something mathematically incorrect or doing something pedagogically unsound. Maybe my students would be difficult to manage that day, and I would be perceived as having poor classroom management (every teacher's nemesis). However, I learned something about myself that I had always suspected. I knew my students well, and I could often assess what had happened in class without even seeing the tape. As a result, viewing the videos became less of an embarrassment and more of a test to see if what I thought had happened in class was accurate. Once past the self-consciousness, I was able to acknowledge my successes and analyze critically lessons without cringing.

Mitch, Eric, Rebekah and I continued to meet once a week to analyze tapes and discuss my teaching. Questions continued to become more mathematically pointed. Why did I do that? What were my teaching goals? What were my mathematical goals? What did I want the students to know? I vacillated between feeling that I had already answered their questions (a million times already—why did they keep asking me?) and secretly knowing that

I was somehow avoiding the real substance of their questions. Could I continue to fake my (perceived) lack of mathematical understanding?

I cannot remember now whether or not there was a specific moment in time when a breakthrough occurred in my ability to feel safe enough to answer more honestly and not feel ashamed by asking questions. I suspect that it happened over time, but I believe that Mitch was largely responsible. His questioning style, while thorough and keen (and sometimes relentless!), was also gentle and sensitive. He made it apparent that he truly cared about the honesty of my answers. No matter what I said, it was valued. He was infinitely patient. I can remember long silences in our conversations when he would ask me to "say a little more" about a feeling or belief. Self-disclosure became not only easier, but also necessary for me. I became fascinated by my inner feelings, and couldn't wait to unearth every memory, detail and revelation. It occurred to me that in the early stages of the project I had feared that I was being rated and judged by my answers, when in reality my open and honest disclosures were essential to the basis of the research. I distinctly remember feeling that by the second summer of inquisition, I was more confident and felt more like an equal participant.

There were several foci that summer of 1996 and while we still continued to fine tune the basis for my beliefs, discussion centered more on my classroom practices and mathematical goals. We began looking at videotapes and writings of other educators. I enjoyed not being the center of attention for a while, and feasted on the rich ideas and practices of Vicki Zack, Deborah Ball and Jim Minstrell. I felt that the relationship among the research team members change somewhat. I thought, "Okay, now you know what I believe, and you've seen me teach for an entire year. How can we make it better?" The desire to improve my teaching became our common mission, and it was very exciting. It was inconsequential that I didn't consider myself a mathematics expert because my mathematical prowess was not the focus. They needed me to help design a teaching plan for the research, and I needed them to achieve my goal of becoming a better teacher. We all had an equal stake in the outcome.

For the next two years, my role as an object of study continued to grow and evolve. I felt increasingly responsible for my contribution to the research. It wasn't just Mitch's research anymore; it was ours. My expertise as the classroom teacher became more and more important as we developed specific lesson plans and activities to engage students in meaningful learning. I had to make decisions about whether an activity was captivating, reasonable in content and length and appropriate to the skill level of my students, and in addition how it addressed my mathematical goals. It became apparent that I had mathematical skills, and that I had really been selling myself short. By the summer of 1998, I felt as though Mitch and I had solidified a true partnership. He had arrived in 1995 with a lot of plans, and he needed a classroom in order to try them out. I had the classroom, and was hungry for new ideas. This collaboration exceeded our expectations. Was it worth the long summers of extra work? Was it worth the discomfort of self-disclosure? Has it changed my teaching practice in a positive way? My answers are yes, yes and yes.

INTO THE CLASSROOM

Earlier, I stated that I had been a successful and well-respected teacher prior to agreeing to participate in the early algebra research project at the University of Colorado. What about my teaching did I want to change? In attending various math workshops and reading articles in journals, it was apparent that one of the main concerns of reform educators and the National Council of Teachers of Mathematics [NCTM] was that students did not seem to have deep understandings of mathematical concepts. Many were able to memorize procedures and do calculations, but were unable to use this knowledge in practical applications or in problem solving. I had seen plenty of this in my own practice, and was eager to make mathematics more understandable even prior to meeting Mitch. I began to look for ways I could tell my students why something was true—ways that I could demonstrate proof. I reasoned that if I could show why, for example, the identity property could generate equivalence, students would better understand the concept of simplifying fractions. Even with repeated demonstrations and examples, I was having limited long-term success with student understanding. I felt I needed help.

During the first two summers of research, I maintained that a core belief of mine was that human beings learn by doing. Not far behind that belief was another one that students often learn from each other—sometimes better than from their teachers. Armed with those tenets, I began calling on students to explain their thinking to others. My uncertainty and insecurity with how to manage this is portrayed in a quote from one of our early classroom tapes in 1995. Inviting a student (all students are referred to by pseudonyms) to the overhead to show her solution to a problem, I asked, "Maggie, would you be willing to just sort of explain yours?" When I reviewed the vague and unconvincing phrase "sort of explain" on the video, I was even more determined to become more proficient in incorporating student presentation and discourse into my practice.

This was when we shifted the focus from me to other educators. I viewed several of Vicki Zack's (1996) classroom videotapes, and began incorporating many of her ideas into my own classroom. Using small groups to facilitate student interaction, I posed problems such as "Washing Hair" (Meyer,

1983), and "Identifying Qualitative Graphs" (van Dyke, 1994), listened to students' ideas as they worked together, and then had students present their solutions to the whole class. I stopped using phrases like "sort of explain" and began to ask students to be clearer about their thinking. I also incorporated more writing as a way for students to explain their reasoning. One problem that worked particularly well for this was from the NCTM (1992) Addenda Series *Geometry in the Middle Grades*. Students are asked to view a series of pictures based on a map. The writing prompt was "Unfortunately the pictures were dropped and got mixed up. Can you put them in the right order? Explain your thinking." Another example was, "How can five pizzas be divided equally among three friends? Explain your thinking using words and pictures." I liked the new direction, and found that some of the most interesting mathematical ideas surfaced and misconceptions were flushed out as I learned how to ask better questions of my students.

I still felt as though I was doing too much talking. I wanted my students to begin asking the questions of each other that I was currently posing. I wanted them to really listen to each other, and to feel safe in speaking to the group. In keeping with my basic tenets of learning from one another and learning mathematics by doing mathematics, it was important to me that my students learned to speak directly with each other. Deborah Ball's (1990) tape on "Shea Numbers" was a true inspiration. After Deborah led some routine class discussion, one of her students posed a theory about odd and even numbers. What ensued in her class was what I had been dreaming of. A student named Shea shared a hypothesis about a mathematical concept, and a spirited discussion followed, led by the students' own arguments, examples and counter-examples. The students were respectful to each other and very engaged. Deborah facilitated this by staying in the background, encouraging different students to share their ideas, and asking good questions that forced students to clarify their opinions. How could I get my classroom to look like that?

The research team discussed the elements of a classroom committed to discourse. We realized that we would need to have conversations with students starting from their first day that this would be a class where they would be expected to talk and listen. We used desk arrangements that allowed all the students to see each other during whole group discussions, but easily converted to an arrangement for small group discussions. We consulted with Lew Romagnano, professor of mathematics education at Metro State College in Denver, about meaningful problems, and questioning and assessment techniques. We selected and developed problems that did not lend themselves to simple calculations, but depended on multistep problem solving and had multiple entry points and solution paths. We also worked on establishing an atmosphere of trust and safety for students to speak publicly to the class, the teacher, as well as members of one's group. To help with this, we asked Laura Till, a mediator and professional facilitator with experience in mathematics education, to be a consultant in the area of group dynamics. We spent two days with Laura doing non-mathematical, trust-building activities that allowed students to share things about their interests and home lives. Then we moved into problem-solving activities with a strong emphasis on hearing about everyone's approaches and students asking each other to explain or elaborate on their ideas. Students also reflected on their experiences collaborating with group members. The adults role-played common class dynamics that inhibited trust and sharing, such as interruptions, poor eye contact, non-constructive feedback, and weak presentations. Students then brainstormed how to correct these staged interactions. We added these contributions to a chart of good listening and speaking skills that was placed in the front of the room and referred to throughout the year. We also revisited these skills with Laura after the winter break.

Then our research group tackled a key issue: What were the big ideas about math that I wanted my students to understand? I had to start thinking about important, general concepts, not just topics or skills. For example, one truth I wanted my students to understand is that mathematics is not mysterious and magical; rather it has order and is logical. I tried to revisit that idea every time they generalized a pattern or made a mathematical connection. A few examples come to mind. The order of mathematics is beautifully illustrated by listing factor pairs of a number. I taught my students to realize that if they listed the pairs in order, they could tell when they had found them all, either by beginning to see repeated factors, or by recognizing that the product of the factors was becoming more "square." Later, during algebra instruction, my students learned that an equation was nothing mysterious because it followed logical steps in pattern generalization: draw a picture, create a table of values, describe the relationships, write an equation, and describe how the equation relates to the picture and the relationships.

Once the big ideas were in place, I could start matching them to relevant units of study and activities. This was a new approach for me. I had never really thought about general mathematical truths while crafting a lesson plan. Teachers are held so accountable for the list of topics that need to be covered, that they focus most of their attention on how to get through the curriculum. I was no exception.

Related to this is the tendency for many teachers to follow a curriculum guide blindly, regardless of what preconceptions and misconceptions their students might bring with them to the classroom about a given topic. Jim Minstrell's (1989) video about his pre-assessment of scientific concepts among his students and subsequent lessons and class discussions was very motivating. I began to take time to expose misconceptions and throw them out to the class for discussion. For example, a few common misconceptions among many sixth graders surface during discussions about division. Many believe that it is impossible to divide a smaller number by a larger one, or if you do, the answer will be negative. Challenging students' deeply held beliefs has led to many spirited discussions, like the one in Deborah Ball's class for which I was so desperate to have take place with my students!

All of these changes came about in the first two and a half years of the study. By the summer of 1998, Mitch and I were collaborating on specific early algebra instruction, which was the big focus of the project. Our meetings had evolved from inquisitions to shared decision-making sessions about how to best sequence and present algebraic concepts to sixth graders. I was in new territory, and I welcomed it. Instead of being embarrassed about something I saw on a video of myself, I was able to take an objective view and ask, "Now, what went wrong there? How could I have facilitated that more effectively?" Mitch and I would brainstorm ideas for revision. We went a step further. Instead of discussing what I was doing or not doing on the video, the focus shifted to what the students were learning or not learning. And instead of focusing on my insecurities, we discussed student learning in a mathematical context.

In the fall of 1998 we put our ideas to work. I was teaching at a middle school in the Rocky Mountain region that was predominantly Caucasian (86 %) with some Asian (6%), Hispanic (5%) and American Indian (2%) students. A small number of students (12%) qualified for free/reduced lunch and special education (13%). The mathematical performance of the students on the California Achievement Test (CAT) had an enormous range, from the 5th to the 99th percentile. During our collaboration, there were typically four to five students in each class who received special education support for their physical and cognitive disabilities. A paraprofessional served the classroom once a week to help meet these students' special needs.

Over the summer we developed a specific curriculum to address early algebraic thinking. The NCTM *Standards* call for algebra learning K–12, where students progress from being able to identify simple patterns to being immersed in formal algebra courses. Based on prior research in this project (e.g., Nathan & Koedinger, 2000b), we believed that middle school students had developed intuitive and informal methods to solve algebraic problems prior to formal algebra instruction, even though these topics were traditionally presented in later grades. However, rather than seeing them expressed in symbolic representations and formal procedures, these intuitive methods tended to be verbal in nature. Our approach focused on eliciting those verbal methods from students in the class, helping them to articulate and abstract them away from particulars of each problem, and then bridging them to more formal approaches. The following lesson on pattern generalization followed a two-week study of writing equations and their inverses to mathematically model written situations such as, "Barb was two years older than Sam." Students had also had some instruction on the order of operations. These topics fit perfectly with the NCTM *Standards* of algebra, patterns, and number and operations, and would extend students' earlier work with equations. I chose two problems that could be solved by discovering a numeric pattern and could also be modeled with objects or drawn on paper. My source was *About Teaching Mathematics, a K–8 Resource* (Burns, 1992). I started with the "Toothpick Building" problem as presented ("How many toothpicks will you need to build a row of 100 triangles?") and then amended "A Row of Squares,"("If you line up 100 square tables in a row, how many people could be seated?") "A Row of Pentagons" and "The Banquet Table Problem" to create my version called "The Dinner Party Problem." The sequence of lessons lasted for approximately three class periods, or a total of 150 minutes.

THE TOOTHPICK BUILDING PROBLEM

On the first day of instruction, I handed the students a copy of "Toothpick Building." I spent some time explaining the pattern on the board before I turned students loose to work with their partners to accomplish two things—finish an incomplete table of values and describe, using words and pictures, the relationship between the number of toothpicks one needs to build a given number of triangles strung together in a row of varying length. My initial goal was for students to get their verbal descriptions of the relationship on the board so that they could see easily each other's thinking and we would have a basis for class discussion.

Our observations with students in previous years (Koedinger & Nathan, 2004; Nathan & Koedinger, 2000a, 2000b) led us to believe that verbal forms of reasoning and representations were natural entry points for students learning to develop skills in algebraic modeling. Prior research also has shown that students have more success using verbal representations to express quantitative relations and solutions than they do with more formal representations including equations and graphs (Koedinger & Nathan, 2004; Nathan & Koedinger, 2000b; Nathan et al., 2002). We drew directly on this evidence when we designed the instructional approaches for these pattern generalization problems, consciously using students' verbal reasoning abilities as a way to bridge to equations (Nathan, 1999).

I was hoping some students would go beyond the recursive method ("you add two toothpicks every time") that requires one to only think about the change of one of the variables (toothpicks) and, instead describe the pattern in terms of both the number of toothpicks and the number of triangles. I circulated among groups while Mitch and Kate Masarik (a doctoral student and research assistant who had joined the project) set up a microphone and video on one pair of students so we could gather data on their thinking. When all pairs of students had put their written explanations on the board, we were excited at the diversity of thought. While several students described the pattern recursively (as expected), there were three descriptions that showed students were attempting to key in on a relationship between both triangles and toothpicks.

A lively discussion ensued where students defended their explanations and looked for similarities and differences. I challenged them to describe how their description was like or unlike the others on the board. Almost every student had some observation to make and I attributed the high level of participation to the public display of all their work on the board. I then asked the students to write an equation that modeled the relationships described by their words. By this time, most had latched onto 2 x number-of-triangles + 1 = number-of-toothpicks as their equation of choice, but I wanted them to be able to test the equation using different values and explain *why* that equation was a reflection of the toothpick pattern. The common summary was "every triangle uses two toothpicks (2 x number-of-triangles) except for the first triangle which has three (+1)." In addition to this, one student was able to generate the inverse relation to find the number of triangles given a number of toothpicks. We were pleased with the students' responses to this first pattern problem and the general pedagogical process that encouraged so much student discourse and reflection. Thus, we were anxious to take their level of understanding one step further.

THE DINNER PARTY PROBLEM

On day two, I had the students for a 90-minute block of time. Once again, Marilyn Burns was my source. I adapted "A Row of Squares" and gave it a context by telling students that I was planning a dinner party for an undetermined number of guests who would sit one person per side at square card tables connected to form a row of tables (see Figure 3.1a). Students continued to work in pairs and were given the charge of completing the table of values (see Figure 3.1b), writing a verbal description of the relationship

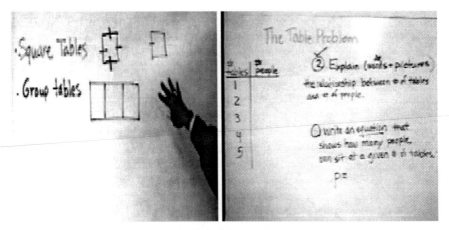

Figure 3.1.   The Table Problem. (a) The arrangement of "guests" at the "dinner tables." (b) The student assignments along with an initial table of values for the pattern showing the number of guests that can sit down if you know the number of tables at the dinner party.

between the number of tables and the number of people that could be seated (using words and pictures) and finally, writing an equation that matched their written description. Once again, we were rewarded with a variety of solution presentations from students.

First, I asked each pair of students to read their verbal description orally and describe any diagrams that they used to understand the pattern. I charged the remainder of the class to ask the presenters for clarification when needed and to ask the presenters how the verbal description or diagram captured the pattern. Thanks to one pair of students, an opportunity arose to explore students' attempts at using relatively meaningless combinations of numbers in one column (number of tables) within the table of values to generate the value in the other column (number of people). As shown in Figure 3.2a, the verbal description "add the number of tables to the number below it then add one to get the number of people" works mathematically, but is completely divorced from the meaning of the variables in the Dinner Party problem. This is because it does not really model the causal or visual structure of the problem, it just uses the successive arrangements of the entries in successive rows that satisfy the pattern. In Figure 3.2b, a student uses hand gestures to show that she combined numbers from adjacent rows to get the number of people. However, this account did not provide insight for algebraically modeling the problem situation.

In contrast, another student (see Figure 3.3) provided a verbal description and diagram that showed how with each successive table only two

add the number of tables to the number below it then add one to get the number of people

Figure 3.2.   (a) A verbal description of the Dinner Party Problem. (b) A student gestures to show how combining values from successive rows in the first column (and then adding 1) gives the value in the second column.

additional people can be seated. The first and last (end) tables, however, each provide seating for three guests. This discussion of students' verbal descriptions of the pattern lasted for about 20 minutes, with most students remaining engaged and on task, but I felt a stretch break was in order before we moved on to their equations.

When students returned from the break, we discussed their mathematical equations. The equations were more varied than were those given in the Toothpick problem. The pair of students from Figure 3.2 showed how their procedure of adding values in successive rows could be written symbolically (see Figure 3.4). They came up with $T + U + 1 = P$, where "U" represents "the number directly *under*" a value in the tables column.

I suggested to these students that three variables (T, U and P) in one equation was confusing, and asked them if they could re-write "U" in terms of the number of tables (T). They could! It was agreed that U always equaled $T + 1$; that is, the number in the row below was the original number plus 1. The new equation became $T + (T + 1) + 1 = P$. Because of earlier equation work, students were able to combine like terms and generate $T + T + 2 = P$. It was then a logical step to produce $2 \times T + 2 = P$. Our discussion included several checkpoints where I asked students to test the equations by inserting values from our chart and confirm the equations were equivalent.

"That's great!" I told them, "but *why* does that equation fit the situation? How does it describe what is happening with tables and people?" Their

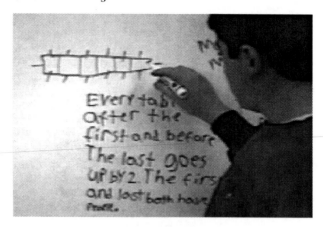

Figure 3.3.   Another student explains how his verbal
description models the dinner party.

conclusions were that there are always two people at the sides of every table
(2 x T) and there is one person at each end (+ 2). This time, more than
one pair of students wrote an inverse equation and tested it for accuracy
(see Figure 3.5). For verification of understanding, I assigned a homework
problem where students had to generalize a pattern and write an equation
for hexagonal tables. Nine out of ten pairs of students were able to accurately describe the new pattern and write an appropriate equation.

In reviewing our tapes of this lesson, we concluded that there had been
a multitude of process standards and principles at work. Our work emphasized the building of new mathematical knowledge through problem solving, particularly by having students apply and adapt strategies and reflect
on their process through extensive communication. Students had to organize their thinking and present it to their peers verbally and in writing,

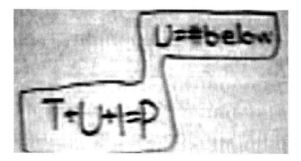

Figure 3.4.   The equation that followed from the
verbal description of Figure 3.2.

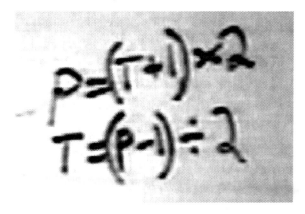

$$P = \left(\frac{T}{2} + 1\right) \times 2$$

$$T = (P - 1) \div 2$$

Figure 3.5.   An example of student work showing the equation that modeled the Dinner Party Problem and its inverse equation.

where they were encouraged and coached to use the language of mathematics to communicate their ideas clearly, coherently and precisely. Their understanding of different representations was greatly enhanced, as witnessed by their more sophisticated use of verbal, spatial and symbolic notation with both pattern problems.

A large impetus for our involvement in lessons such as these was to address greater participation and involvement of students across the achievement spectrum. In looking back through our tapes, we discovered that *every* student in the class either wrote their thinking on the board or participated in the class discussion. We wanted to elicit students' intuitions and invented strategies that were clearly meaningful for them, and chose the partner/presentation model because of what we knew and believed about sixth graders as learners—they like to use visual representations and they like to talk to each other. This related directly to my core beliefs that were unveiled during our earlier summers of "teacher interrogation," where I stated that I felt children learn by doing and that students learn from each other. Our model of instruction also included frequent, diagnostic assessment. In fact, it was woven into the very fabric of the lesson planning, as we continually moved back and forth between pre-assessment, class discourse about student approaches, our verbally based theory of student mathematical reasoning and development, and the continual refinement of curriculum activities and instructional practices.

Despite the success of these lessons (as observed later on the video and by the students' positive performance on the post-assessment), I experienced some of my old fears and insecurities while teaching, particularly when students would invent equations that I had not anticipated. It was occasionally unnerving to feel "on the spot," trying to analyze an equation quickly, believing   I should understand instantly everything students were generating. Sometimes when I asked a student, "How does your equation work?" I was not sure of the answer myself, and once again would feel little prickles of inadequacy. But when I studied the classroom videotape later in the week, I felt that my questioning and teaching techniques used to help students flush out their ideas were very effective. For example, I frequently invited multiple responses to a question, asking with great frequency if there were other ideas or approaches. I consciously increased my wait time. I looked for occasions where I thought students would benefit from working things out on their own, or talking briefly to a neighbor. After just a few minutes into a new problem or activity I would often asked students to share with the whole class their initial ideas of how to start. I tried to foster the view that solutions did not speak for themselves, but needed explanations to be complete. I encouraged students to ask their peers questions before coming to me. I also asked students to wait until a person was done speaking before raising their hands, in order to give students the sense that they had the floor and did not have to rush their thinking. Finally, if I were uncertain about a student's idea or comment, I would always ask for clarification, or invite the student to show it on the board. Students had long, engaging discussions. They generated sophisticated equations. They were subsequently able to take their strategies from these very visual problems and apply them to more abstract patterns that did not have a context.

Prior to my participation in this research project, I am certain I would have approached this topic differently. First of all, I would not have spent the time to set up a classroom environment where students felt as free to explain their thinking. The preliminary work setting up classroom norms made students feel safe when asked to speak to me, their classmates, to Mitch, or into the microphone. Secondly, I probably would have guided students more in an effort to get to the heart of the generalization more quickly. We would not have taken the time to dissect the different equations, particularly if they did not work. I believe I would have given more direct instruction in the use of variables, instead of letting the concept evolve naturally. In general, I would have gone faster, done more talking, and not given my students as much credit for their innovative solutions, in an effort to cover the topic with greater efficiency.

EPILOGUE

Each of us, the researcher and the teacher, learned a great deal from this collaboration. From close and frequent observations of students, Mitch learned that students do indeed have powerful intuitions about mathematics that are often overlooked during mathematics instruction. Even prior to instruction, students as young as sixth grade can reason well about algebraic relationships and can articulate these solutions when given the opportunity. Students may not be aware of the power and relevance of these methods, and may think that they did not do anything, or that they cheated by using methods such as guess and test.

From close collaborations and observations of a couple of teachers, Mitch learned that teachers face amazing challenges as they strive to satisfy the constraints imposed by curricular demands (including those imposed by standards), the needs of students, and the inflexible time demands of the school day and year. Mitch learned that teachers could be driven by overarching goals and tenets, such as valuing the ideas and opinions of everyone in class and even their own experiences as middle school students learning the mathematics for the first time. Mitch found that middle school mathematics teachers face unique challenges, since they are asked to take on more and more advanced areas of instruction (algebra is commonplace now in middle school though it was not so when we started our collaboration), though they often have not had a great deal of mathematics preparation (most he met were licensed as elementary educators, and had insecurities about middle level mathematics). But many middle school teachers were highly motivated to enhance their teaching practices and their understanding of mathematics, and welcomed professional development opportunities that provided greater content knowledge and occasions to reflect on their teaching practices.

For me, (Amy) the experience of being a classroom teacher involved in a research project has been rewarding, but not always easy. It was very difficult to admit that despite my sterling reputation, I needed to improve. It was unnerving to know that I was being taped doing a lesson I had never tried before, full of mathematical concepts that in the not-so-distant past had been daunting for me. Presently, it is even harder than ever to get through my curriculum, because I am so committed to my students understanding key ideas in great depth. Three years after the completion of this study, I changed schools. I am currently working at a very different middle school (still in the Rocky Mountain region). The school is almost three times larger, and has a greater percentage of minority, low-income, and second language students, with 30% Hispanic, 20% English-language learners, and 32% receiving free/reduced lunch. The current political climate, with its focus on teacher accountability and high-stakes testing, is predominantly oriented toward standardized test score gains. Too many of my current students have scored "unsatisfactory" on state mandated tests. Because of this, I am worried constantly about time and coverage. How can I possibly cover all they need to know in order to make positive growth on these assessments? I worry that my students will not make adequate gains and that future teachers will see gaps in their knowledge. It is time consuming to plan lessons that involve small group tasks and meaningful discourse. Fostering the discourse in class is a constant struggle. Kids are only used to talking to each other socially; they resist truly listening to a peer's academic thoughts. It is difficult for the second language learners to participate without a lot of scaffolding of the language beforehand. I really have to be prepared prior to class so that I can anticipate the possible directions the lesson will take. And during the lessons themselves, I cannot rest. I need to pose good questions, listen to students, challenge them to be clearer in their thinking, and keep the mathematics visible. It is exhausting.

However, as I look back on how I have evolved as a teacher and a learner, I know that as a result of this experience, I am a more confident and accomplished teacher. Instead of viewing the mathematical community with anxiety and feeling uncertain about my abilities, I am developing a critical eye and ear for authentic mathematical tasks. My goals for my students are much clearer, and what they master conceptually has become more important to me than the list of required topics. My students have opportunities to test out their ideas, engage in mathematical discussions, and present their findings in an environment that values their opinions and insights. But the most beneficial outcome has been the opportunity to engage in a high level of self-reflection. I agreed to participate in the original research project because I believed it would allow me to reflect on aspects of my instructional practices. In particular, I hoped to gain insights into students' mathematical intuitions about algebraic reasoning prior to formal instruction. I have learned more about teaching from this experience than any post-graduate class or in-service training. To have the time to truly think about one's core beliefs and subsequent practice, to have conversations with colleagues and professionals who value those beliefs, and then to successfully put these ideas into practice—this has been life changing.

ACKNOWLEDGMENTS

The authors gratefully acknowledge the assistance, wisdom and inspiration of the following people: Rebekah Elliott, Eric Knuth, Kate Masarik, Lew Romagnano, Laura Till and Isobel Stevenson. The research discussed in this chapter was funded by the James S. McDonnell Foundation CSEP Program under the "Bridges for Representational Fluency" grant to Kenneth R. Koedinger, Mitchell J. Nathan, and Martha Wagner Alibali.

REFERENCES

Ball, D. (1990). *Shea numbers* [Motion Picture]. (Available from Mathematics and Teaching through Hypermedia, University of Michigan School of Education.).

Burns, M. (1992). *About teaching mathematics: A K–8 resource.* Sausalito, CA: Math Solutions Publications.

Koedinger, K. R., & Nathan, M. J. (2004). The real story behind story problems: Effects of representations on quantitative reasoning. *Journal of the Learning Sciences, 13*(2), 129-164.

Meyer, C., & Sallee, T. (1983). *Make it simpler.* Palo Alto, CA: Dale Seymour Publications.

Minstrell, J. (1989). *James Minstrell's teaching tape* [Motion Picture]. Seattle, WA: Talaria.

Nathan, M. J., & Knuth, E. (2003). A study of whole classroom mathematical discourse and teacher change. *Cognition and Instruction, 21*(2), 175-207.

Nathan, M. J., & Koedinger, K. R. (2000a). Moving beyond teachers' intuitive beliefs about algebra learning. *Mathematics Teacher, 93*, 218-223.

Nathan, M. J., & Koedinger, K. R. (2000b). Teachers' and researchers' beliefs about the development of algebraic reasoning. *Journal for Research in Mathematics Education, 31*, 168-190.

Nathan, M. J., & Koedinger, K. R. (2000c). An investigation of teachers' beliefs of students' algebra development. *Cognition and Instruction, 18*(2), 209-237.

Nathan, M. J., Stephens, A. C., Masarik, D. K., Alibali, M. W., & Koedinger, K. R. (2002). Representational fluency in middle school: A classroom based study. In D. Mewborn, P. Sztajn, D. White, H. Wiegel, R. Bryant, & K. Nooney (Eds.), *Proceedings of the twenty-fourth annual meeting of the North American chapter of the International Group for the Psychology of Mathematics Education.* Columbus, OH: ERIC Clearinghouse for Science, Mathematics, and Environmental Education.

Nathan, M. J. (1999 April). *An instructional theory for early algebra that incorporates research on student thinking, teacher beliefs, and classroom interactions.* Paper presented at the Annual Meeting of the American Educational Research Association, Montreal.

Richardson, V. (1994). *Teacher change and the staff development process: A case in reading instruction.* New York: Teachers' College Press.

van Dyke, F. (1994). Activities: Relating to graphs in introductory algebra. *Mathematics Teacher, 87*(6), 427-432, 438-439.

Zack, V. (1996). *Group of four: Washing hair* [VHS] (Videos taken by Author).

CHAPTER 4

LEARNING TO TEACH

Alf Coles
Kingsfield School, South Gloucestershire, UK

Laurinda Brown
University of Bristol, Bristol, UK

The overt product of research is some supported assertion(s). A covert product of research is a transformation in the perspective and thinking of the researcher. Undoubtedly, one of the most significant effects of any piece of research in education is the change that takes place in the researcher. (Mason, 1996, p. 58)

ALF: INTRODUCTION

When I think about my development as a researcher, the journey is inseparable from learning to teach. The changes that have taken place in the time I have been involved in research have been in my classroom practice. I teach in a co-educational state-funded school for 11–18 year old students on the outskirts of the city of Bristol, UK. The catchment area includes communities ranked in the lowest third in the UK on scales of deprivation and the school is poorly funded compared to UK norms. It has an intake of

Teachers Engaged in Research
Inquiry Into Mathematics Classrooms, Grades 6–8, pages 55–80
Copyright © 2006 by Information Age Publishing

approximately 230 students a year, whose prior attainment is significantly below the national average. And, while most students make good progress in the school, by the age of 16 overall results on external examinations are also below national averages. First-year students (aged 11) are taught in mixed ability groups in mathematics.

I see myself as developing communities of inquiry within my classroom, where students are motivated to work on their own questions related to mathematics. Other teachers have been motivated to develop their practice after seeing these students working in mathematics lessons. I convene a cross-disciplinary classroom research group within my school where we share teaching strategies to support students' critical thinking skills and I lead the mathematics department in my school where our agreed aims include:

> To create communities of inquiry into mathematics in our class-rooms, in which every individual can be heard, have their views respected and grow in self-esteem. ... For students to find and develop their 'mathematical voice' in making meaning of the situations that we present them. ... For all of us (staff and students) to be reflective about our experiences and to learn from them.

I also see myself as facilitating a community of inquiry within the department, where teachers are motivated to work on their own questions and issues related to the teaching of mathematics. In this chapter, I describe the journey of my development that has led to the convictions that allow me to work with students and adults in some of the ways described above.

ALF: STARTING TO TEACH

I began teaching in England in 1994. During my one-year postgraduate education program, I had wanted to establish a humane classroom in which students expressed their autonomy and were engaged in self-motivated and creative activity. I had been influenced by ideas such as those in *How Children Fail* (Holt, 1964) and I wrote the following in 1995:

> July spelt the end of my first year of teaching. In the course of the year, I gained a sharpened image of how I would like to be in class, and of the ways in which I would like my students to be working. Along with this clarification, however, came a widening gap between the image and what actually was happening. Ideally, I see myself being

matter-of-fact, avoiding condescension and blame and even praise; capturing attention and emotion and directing this into students' own awareness. In reality, I rarely held the classes' attention. I realize I did condescend to my first years and … often became negative.

I would like my students to feel that Maths was something that connected with them, that its roots lay in their own intuitions, and that it can be tackled independently. I hoped the study of Maths would help students to realize the potential and power of their minds. None of this happened. Indeed, many students now lack the faith that there is much to be gained from Maths at all. Far from opening minds, it is clear that some students have a diminished sense of their capabilities and an impoverished self-esteem. The fundamental features of what I envisaged in my classroom have not materialized. (Diary, Alf, 07/95)

I found it difficult to use these realizations positively. They engendered despair. My beliefs about what I wanted in a classroom were not seemingly linked to any actions that might help realize them. Before the end of the summer term, in my first year of teaching, I met Laurinda Brown.

LAURINDA: INTERLUDE 1

I work as a lecturer in mathematics education in the postgraduate teacher training program at the University of Bristol, UK. I had been researching my own practices and, at the time that I met Alf, had been looking for a teacher, who had not already worked with me and who was finding teaching a struggle, to work with to develop my theories-in-action. This is Alf's story and part of that journey was meeting a teacher-educator who was interested in his development and his questions. I brought with me the idea of *purposes* (Brown & Coles, 2000) as a language for us to talk about teaching. Purposes were not philosophical statements such as "autonomy" nor were they particular teaching techniques. They were in a middle position, describing practices or intentions such as "How do I know what they (the students) can do?" These purposes are identified through reflections on practice, or telling the stories of events through the details that stay with you allowing the question or issue raised to find a name. I was interested in whether we would be able to find an initial purpose to work with. The important point is that this purpose had to be identified by Alf, so that he would be able to start looking for teaching strategies to continue to develop that aspect of his teaching.

ALF: SILENCE

Laurinda encouraged me to look back over my first year of teaching and, if no whole lessons were any good, maybe describe some *moments* when I approached my ideal image of how I would like to be in class. Two stories initially came to mind.

- Story 1: During a lesson (to 16–17 year olds) on partial fractions I was going through an example on the board, trying to prompt suggestions for what I should write. Some discussion ensued among the students, which ended in disagreement about what the next line should be. I said I would not write anything until there was a unanimous opinion. This started further talk and a resolution among themselves of the disagreement. I then continued with the rule of waiting for agreement before writing the next line on the board.
- Story 2: Working on significant digits with a year 9 class (students aged 14), I wrote up a list of numbers and got the class to round them to the nearest hundred or tenth, ... Keeping silent, I wrote, next to their answers, how many significant digits they had used in their rounding. Different explanations for what I was doing were quickly formed and a discussion followed about what significant digits were.

I was driving at the time of recalling these stories and without any prompting said, with energy: "It's silence, isn't it? It's silence." My awareness of *silence* being a sameness in these stories felt like it came from staying with the detail of the accounts and seeing a pattern that was there. Laurinda and I had found an initial focus for our work together. We planned jointly lessons that would begin with my own silence. I had something in mind (i.e., my own deliberate silence), which could inform my planning and actions. There was a markedly more engaged reaction from some of my classes. For almost the first time I had a real conviction about my actions in a classroom, linked to my ideals about what I thought was important in teaching.

It seems significant that silence is something I recognized and valued in other contexts. I wrote around the time of the incident above, about my own story of silence:

Silence is also in waiting. I think I learned how to wait in Zimbabwe—hours by the side of roads waiting for buses and lifts. I have an image of waiting with Zimbabwean friends who would just sit ...

A few months after working on silence as a purpose in my teaching I wrote:

> One discipline that has also come out of the work with Laurinda is that of *staying with the story.* In my notes on teaching in the first year, the observations are in general distant about whole classes, with observation and analysis all mixed in. What I have been working on this year is forcing myself to hold back the analysis and stay just with stories about individuals or groups of individuals. Analysis (or synthesis) from these data then has the possibility of throwing up something I had not been aware of before. There was previously something about the mixing of analysis and observations that meant that I was never surprised—everything that happened was seen through the filter of what I expected. There was little chance of my accessing those things I did that I was unaware of, but which yet had profound effects. (Diary, Alf, 10/95)

This extract is from a diary of critical incidents and reflections I started to keep and that I now see as the beginning of my own research journey.

LAURINDA: INTERLUDE 2

Alf and I became co-teachers and co-researchers rapidly. The questions about practice that we worked on remained his own and I was researching his development as a teacher in relation to my concept of purposes. However, through jointly planning for lessons for which I would be in the classroom, we could swap the teacher's chalk and the researcher's pen when something arose that we had planned for. This was a rich situation and meant that Alf could observe his own students in his own classroom reacting differently and we both had access to his observations of teaching to support the reflections. The separation of detailed observation from analysis continued to be a useful research tool and also helped Alf to hold off from judging lessons, and consequently himself, as not meeting his ideal.

What, perhaps inevitably, happened was the development of theories in relation to purposes within the teaching. Bateson's (2000) (which I originally read in 1972!) work was influential for me and, as Alf started to read this writing, became so for him too. Our joint reflection on his developing teaching led to an idea of what we have come to call *metacommenting.* Metacomments are statements by Alf describing behaviors in his students that he likes because they fit with his ideas of them *becoming mathematicians.* The culture of the classroom develops through these metacomments (e.g.,

"Josie and Chris are getting organized;" "Why? is a good mathematical question to ask.") and patterns in the students' behavior emerge. We have noticed that, as time passes, students themselves comment on behaviors (to each other or in their writing) that they find useful such as "Try to ask a question when you're stuck." Not all the behaviors metacommented upon by Alf become established in the classroom patterns of behavior. In my formal lesson observation notes some metacomments only appear once. Also, as I observe the developing cultures in different classes, although there are obvious similarities, each established culture emerges from the interactions of particular communications and so is different since different behaviors have been seen and commented upon. What is observable is that some of the metacomments from Alf, and the students themselves, become purposes for the students that support their mathematical development.

Out of this developing journey of becoming a teacher, we were also theory-building as collaborative researchers about the teaching and learning of mathematics and the development of these classroom cultures.

ALF: FINDING A PURPOSE

As Laurinda and I continued to work together during my second year of teaching, we identified other foci, or purposes, that I used to inform what I did in my classroom (e.g., "How can I structure this lesson so that I will not be needed by the students?", "It's not the answer that's wrong.", "sharing responses", "getting organized"). These labels are fairly meaningless for anyone else, but for me, focusing on such purposes gave me conviction in my planning and teaching. I was offering this activity or this way of working for a reason that was linked directly to what I wanted to achieve in a classroom.

Laurinda and I would always work together on the mathematics of the problem or activity that we offered. Through doing this I became more aware of what mathematics I was using or needed. The process of working on the mathematics for ourselves at our own levels also seemed crucial in terms of identifying rich questions or starting points. I felt this process, particularly in identifying a range of possible responses to any starting point, helped sensitize me to what students might say or do and helped me develop some flexibility in terms of my own responses to what students did. We used a phrase—"planning to conviction"—that, for me, was about working to a point of being sure there was enough that was interesting mathematically within my starting point and also having some image of how the students might respond and what I might do next.

In recognizing the power of having some overarching sense of what I was doing that could inform my actions and responses, I began to work on ways of offering students a purpose for a sequence of lessons that might help similarly to inform their decision making. For example:

- I had observed Laurinda while she was teaching one of my classes (of 12–13 year olds), tell the students that their focus (purpose) for this lesson was "not so much on the actual finding of rules but on developing strategies for finding rules."
- We planned a series of lessons on drawing graphs of rules during which I issued the challenge (which became a purpose for the students): "In five lessons' time I'm going to come into the room and write up a rule. Your challenge is to be able to tell me what the graph would look like without having to plot it." This idea is one that I found could be adapted to different situations.
- Working on area, leading to the Pythagoras Theorem, students are given grid paper with dots at each vertex of the grid. They are invited to draw squares with the rule that all the corners of the squares must be on dots of the paper. I issue the challenge: "Find a square with an area of 10." Purposes that usually develop out of this challenge are: "What areas of squares can we get?" and "How can we be sure we've got them all?"
- Write down a three-digit number. Write down the same number with digits reversed. Take the smaller away from the larger. Now take this answer, write it down with digits reversed and add these last two numbers. Many students find they get the total 1089. I issue the challenge: "Can you find one that doesn't work?" This can become a purpose for many students. Often, some students become quickly convinced that the challenge is impossible and they or I might offer the question/purpose: "Why does it always work?"

In lessons in which I was clear about the purpose for the students I found I no longer needed to direct everything that the students did. I was able to encourage students to make their own decisions about what to try out or look at, within the limits imposed by the challenge. At the end of my second year of teaching I wrote:

I have experienced, this year, classes working in an autonomous and motivated way—coming to important realizations—over a series of lessons. This is by no means the norm but I feel it could be. I have become aware of my awareness of what energy students have. When six students came up and showed me their homework before the beginning of a class, I was able to see this not just as a distraction to what I was wanting to do but as an indication that there is perhaps something here that we can all work on. (Diary, Alf, 10/95)

I was beginning to be able to reflect on and develop my practice. Put in other words, I had begun to actively research my practice. Seeing the beneficial effects in my classroom, I was motivated to continue some kind of systematic reflection.

ALF: FINDING AN ISSUE

In my third year of teaching, I began a Masters in Mathematics Education (M.Ed.) program at the University of Bristol, Graduate School of Education, while Laurinda began weekly visits to my school to observe and sometimes co-teach my classes for a day. The explicit aim of the Master's program is to support teachers in researching their own practice of teaching. This involves, initially, choosing an issue in your teaching that you want to work on and develop. I became interested in the issue of algebra and in ways of sustaining work on algebra with students. One seed of this interest was my awareness that, in the lesson I saw Laurinda teach my class (of 12–13 year olds), the algebraic activity of the students had seemed very natural—a contrast to my own experiences of teaching. My understanding of what algebra was broadened significantly. I associated initially algebra with the use and manipulation of letters. Through readings and conversations, I began to see algebra more as a way of thinking, which always seemed to entail some sort of stepping back from whatever process you are involved with, to become aware of the process itself. I was influenced by Arcavi (1994):

> Algebraic symbolism should be introduced from the very beginning in situations in which students can appreciate how empowering symbols can be in expressing generalities and justifications of arithmetical phenomena ... in tasks of this nature, manipulations are at the service of structure and meanings. (p. 33)

Laurinda and I began to find the following definition of algebra the most useful in our conversations and observations. Algebraic activity involves:

(a) Generational activities—discovering algebraic expressions and equations;

(b) Transformational rule-based activities—manipulating and simplifying algebraic expressions, solving equations, studying equivalence and form (adapted from Kieran, quoted in Royal Society/Joint Mathematical Council [RS/JMC], 1997, p. 8); and

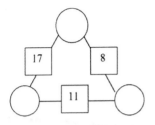

Challenge:
Place numbers in the circles so that the number in each square is the sum of the numbers in adjacent circles.
Is a solution possible? If so, is it unique?
What constraints are there on the numbers that can go in the squares so solutions are possible?

Figure 4.1.   An example of an arithmogon puzzle.

(c) Global, meta-level activities which involve awareness of mathematical structure, awareness of constraints of the problem situation, anticipation and working backwards, problem-solving, explaining and justifying (Kieran, quoted in RS/JMC, 1997, p. 12).

The test of any research idea or definition is, for me, whether it allows me to see more in my classroom or see what is happening in a different light. I was interested initially in exploring what my students did when they were thinking algebraically and whether I could distinguish the three components above in practice.

In one M.Ed. assignment, I interviewed two pairs of high-achieving students, one pair of 15-year-old students and one pair of 18-year-old students, as they worked on a problem. The problem is sometimes called arithmogons (McIntosh & Quadling, 1975) (see Figure 4.1).

This problem could be tackled algebraically. The older students first explored the problem numerically (generational activity), until they had some sense of what was going on, and then moved effectively to an algebraic representation and solution. At the point where these students gained an insight into the structure of the problem (global, meta-level activity) that lent itself to the use of symbols, they were able to introduce letters, derive an equation, solve it (transformational activity) and relate their solution back to the problem (global, meta-level activity). The 15-year-old students, on the other hand, reached for the symbolism quickly and introduced letters, deriving a number of different equations that they tried to manipulate (transformational activity), but they became bogged down in the transformational work and lost contact with what their equations told them.

The 18-year-old students were operating with a much greater sense of control. They exhibited both generational and global, meta-level activity that the 15-year-old students did not, and this seemed decisive in them being able to solve the problem. There was evidence that the 18-year-old students knew what they were looking for when they were exploring the problem initially, since they continued this exploration just until they gained the insight they needed. This study also gave me evidence that the definition of algebra was a useful one and that what I needed to work on in my teaching was supporting students in global, meta-level activity. This experience led to asking the question "Would it be possible to create a classroom culture of *becoming a mathematician* with 11-year-old students so that when they themselves were aged 15 they would be operating as the 18-year-old students did?"

ALF: RESEARCHING ALGEBRA

Laurinda and I started work in September 1998 (my fourth year of teaching) on a project, funded by the UK government's Teacher Training Agency (TTA), to investigate the question above with one mixed-ability class of 11-12 year-old-students (year 7) that I taught. This class had a wide range of prior attainment, from two students with recognized learning difficulties to several students who had well-above-average scores on standard tests. The class as a whole had an average attainment below the national norm.

I decided to offer the class the challenge, or purpose, for the year of *becoming a mathematician*. A large part, for me, of becoming a mathematician is learning about thinking algebraically and using algebra. We hoped the students would find a *need* for algebra by working on activities that required algebraic thinking to make progress.

The evidence at the end of the TTA project was that students did develop a need for algebra within the culture of the classroom over the course of one year of teaching. Evidence for this study was collected by the use of: (a) tracking development in students' written work—including asking them to write at the end of each topic on "What have I learned" within the mathematics and about being a mathematician, (b) video recordings of lessons every two weeks, (c) lesson observation notes by Laurinda, and (d) my post-lesson write-ups. I also conducted semi-structured interviews that were audio recorded with three pairs of students at the start and the end of the first term. The data informed the regular discussions Laurinda and I had about the class. We would, for example, transcribe sections of video or audio recordings of incidents that seemed significant (e.g., Coles & Brown,

2000). We were looking for patterns in the data, in particular in relation to the students' algebraic activity (as defined in the previous section).

The interviews were useful particularly in terms of tracking students' developing use of algebra. The three pairs of students were chosen to reflect the attainment range of the class. They were asked in the first interview (about three weeks into the term) to talk about differences between mathematics now and at primary school. The students reported that before arriving at secondary school they had either little or no exposure to algebra or ideas of proof. The students were given a numerical problem that lent itself to an algebraic solution (the arithmogons problem from Figure 4.1). There were striking improvements in the sophistication of all the students' approach to these problems from the interviews at the start of the term and at the end. For example, at the end of the term one pair was able to move from trying a few numbers in the problem to deriving a rigorous algebraic proof of one aspect of the task, which they were able to fully justify (demonstrating generational, transformational and global, meta-level algebraic activity) (see Figure 4.2). At the start of the term this pair had tried out different numbers and shown no evidence of further algebraic activity.

The final interviews and students' exercise books provided evidence that by the end of one term almost all students were asking their own questions within problem situations.

Several students were able to use algebra in order to answer their own questions. The example in Figure 4.3 of using algebra for proof is from the book of a student of average attainment relative to the class. She was given this number trick: "Think of a number. Add on one more than that number. Add nine. Divide by two. Take away your original number."

"What theories are there? Are there any patterns?"

"Can I predict what the answer is going to be?"

"All my answers come to six; why?"

"I wonder why some numbers go in one column and other digits go in another?"

"I wanted to see what it would come to with four on each side."

"Can I start to guess how many answers there will be?"

Figure 4.2.   Examples of asking questions from six students' books
(across the attainment range)

```
                        N
          (+N+1)     2N+1
          (+9)       2N+10
          (+2)       1N+5
          (÷N)       5


                 it will always end
                   up with 5
```

Figure 4.3.   Extract from student's exercise book

There is evidence shown in Figure 4.3 that the student naturally per-formed algebraic manipulations. She had had no drilling in techniques and yet within the context of the problem she was able to perform opera-tions that are often seen as complex and difficult to teach. There was no explicit invitation to use algebra. The student had her own question: "Will it always end up 5?" and recognized for herself the power of an algebraic approach to answer it. (For further details and examples, see Coles and Brown, 2000.)

In lessons, students asked with increasing regularity in problem solving situations: "Can we do this for N?" In working on these situations, over half of the students showed evidence of using algebra to express their ideas. Across the attainment range, several students were explicit in their written work that they had recognized two different uses of algebra: (a) to prove a result which they believed to be true, and (b) to show the workings of a

- "I've learnt mathematicians have to think quickly and solve lots of problems. You've got to jot little theories down. On a lot of theories you have to write why ... You've got to confirm things (e.g., abc). You got to explain your findings."
- "I have learnt that mathematicians think about why things go in order and why things add up to that number and can I predict this?"
- "... mathematicians are always thinking how does it work and why ... to find out if it works you should use N and carry on from there."

Figure 4.4.   Extracts from students' books, writing about
"What I have learnt" in mathematics

problem (i.e., the use of algebra to express a rule). Students did not view finding a rule or proof as the endpoint of problems but as a prompt for further activity. This result was clear from the students' exercise books.

All students showed an understanding of the meaning of algebraic statements in different contexts. One example of this comes from an activity we call "Functions and Graphs." As part of this activity, students choose functions such as, N -> 2N+1, substitute the numbers 5 to –5 (e.g.., 5 -> 11, 4 -> 9) then plot these pairs of numbers as coordinates to create a graph. The challenge is to predict what the graph will look like just from the rule. All students were able to write and interpret such functions algebraically, substitute numbers, create graphs and comment on what they noticed.

ALF: CREATING THE CLASSROOM CULTURE

Given that it was possible for students to find a *need* for algebra within the classroom culture of that year 7 (11-12 years) class, the TTA project also aimed to describe significant factors in the establishment of such a classroom. Laurinda and I observed the following four teaching strategies to be common in the lessons. These strategies seemed key in setting up a classroom culture in which students asked their own questions and hence found a need for algebra.

1. Giving the students a *purpose* for the year of *becoming a mathematician*

In the first lesson of the year I told the students that this year was about becoming a mathematician and that this meant, among other things: thinking for yourself, noticing what you are doing, asking why things work, looking for pattern. One effect of this purpose was to support the students in being aware of what they did in mathematics lessons by allowing them, and me, to question whether something they did was mathematical or not.

2. Highlighting and *metacommenting* on examples of mathematical behavior, creating labels for students and teacher

In every lesson I aimed, as much as possible, to highlight to students examples of when they or someone else was *being mathematical.* For example, if a student decided to approach a problem in a systematic manner, I metacommented on this by saying something like, "That's an excellent example of getting organized, which is part of thinking mathematically." Over time, phrases such as "getting organized" became labels that students used both in their writing and in talking about what they were doing. "Getting organized" or "asking why" became what students did in mathematics lessons. These labels were part of the culture of the group. Some labels originated from the students. For example, what being a mathematician means is that "it's okay to make mistakes" and that you "share your problems with other people."

   3. The *choice of activities* and teaching strategies used within those activities (e.g., *common boards*)

The study took it as axiomatic that if students were to have the opportunity to experience getting organized, for example, then there must be the space and opportunity for them to have *not* been organized. This had some implications for the ways in which classroom activities were structured. The activities were accessible to all and capable of being extended through question posing. I helped organize the curriculum for the first-year students at my school around seven such activities, which were chosen to cover the content required in the UK's National Curriculum. Each activity would typically last for around ten hours of lessons, so the seven tasks take up over half of students' time in mathematics in the first year. Teachers use the rest of the time to focus on skills or content areas that, through work on the activities, they assess need further practice. One example of such an activity, "Functions and Graphs" is described briefly above. In this activity students will typically be working on (among other things) plotting co-ordinates, reading algebraic expressions, substituting numbers into formulae, manipulating negatives, finding gradients and intercepts of lines, predicting where lines will go, and drawing quadratic functions. In each activity, we aim to make available as many topic areas as we can within one meaningful context. It is partly this complexity that allows students to ask their own questions and enables teachers to spend so long on each activity (see www.mathsfilms.co.uk for a more extensive write-up of "Functions and Graphs" and other activities).

   In each activity, I put forward overarching questions or principles within which I encouraged students to generate their own work and ideas. Self-checking mechanisms were put in place so that I did not have to be the arbiter of whether students were right or wrong. For example, a "questions board" and a "conjectures board" were set up which students wrote on, so that their ideas could contribute to the work of their peers and stimulate class discussion. In the "Functions and Graphs" activity, students drew graphs on paper and pinned them to a board. From this display we would discuss questions and conjectures that were recorded on the other boards. Some examples of students' conjectures are: "Rules with the same number in front of N will be parallel.", "The number you add at the end tells you where the graph crosses on the y-axis.", "If a rule has two N's in, it will be curved." A copy of such a *common board* from a different activity is shown in Figure 4.5.

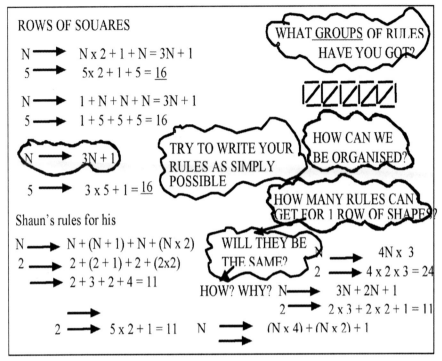

Figure 4.5.   An example of a *common board*

Another activity we call "Pick's Theorem." This involves students drawing shapes whose corners lie on the nodes of a square lattice and recording the number of dots inside (I) the shape, on the edge (E) of the shape, and the area (A).

The two shapes in Figure 6 are classified as 8-dot shapes (since E + I =8). Students investigated what other 8-dot shapes they could find and came up with conjectures, such as "With an 8-dot shape, if I = 0 then A = 3." A board was set up for students to draw up any shapes for which they could not find the area. Other students treated these as challenges. Tables of results were also recorded in a common display area. The overall challenge was to be able to predict the area of any shape just from knowing the numbers of dots both inside and on the edge.

The common boards provided a mechanism not only for students to check their own and others' work but also to allow lower-achieving students access to accurate results or observations that had been generated by the class. It was by the action of classifying such information that it became natural for students, across the attainment range, to ask their own questions.

It was important for me to recognize when to introduce skills to the students and allow them time to practice them. The need for skills such as

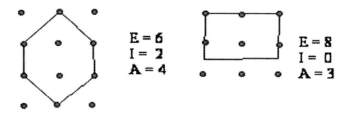

Figure 4.6.   Two shapes drawn on a square lattice grid

basic arithmetic or finding areas or solving equations arose from the problem situations the class were dealing with. For example, working on 'Pick's
Theorem' it was clear many students were not able to find the area of a triangle without counting squares. In one lesson we had a focus on this skill,
which students needed. I invited a discussion of how different students
found areas of triangles (without counting squares) and got students to
practice the different methods they came up with. They then used these
skills to find the areas of the more complex shapes they had drawn as part
of the Pick's Theorem problem. In this and other instances the students
were given the opportunity to work on skills within meaningful contexts.

   4. Emphasizing *students' writing*

   From the first lesson of the term, I encouraged students to write down
what they were doing, what question they were working on and what they
noticed or found out. This was established as being part of becoming a
mathematician. The effect of this was to provoke in students further
awareness of what they were doing, and help them to make links with what
I took algebra to be about. At the end of each activity I got the students to
write about what they had learned—both about new mathematical skills
they had developed and about what they had learned about the process of
thinking mathematically.

ALF: RESEARCH RESULTS IN USE IN OTHER CONTEXTS

A teacher of English at my school came to observe a lesson with my year 7
class during the year of the TTA research, when the students were using a
common board. Partly due to her interest, I started and ran a cross-curricular group of teachers looking at teaching strategies that can be used across
subjects to promote student involvement, choice and independent thinking. A number of teachers in this group were able to adapt the mechanism

of common boards to their own subject. The text below was spoken by a teacher of German, Nick Sansom, who is part of the group, during a taped conversation with Laurinda:

> Well, one of the things I've done recently ... it was a vocabulary build-ing exercise we're learning about ... one of the topics is what they do in their free time. And the main verbs that are used are things that they play and places that they go. So you've got "to play" and "to go" ... And I let them, with the dictionaries, start looking through and finding things and using their imagination, what they play, where they go, find the words. Then they come up and write them on the board. And if another group had the same thing then they compared the spellings, see if it was right. So the end of the activity, we'd go through and select and ask if there were any problems with the spell-ing, and, have you got this, say, are there any differences, ask "Who's right?" then cross reference with someone else. So that worked out quite well in terms of building vocabulary ... they ended up writing up a communal list or adding bits to their own lists from the list on the board. And I think they may well have got more information than they would normally have had.

The single idea, in the cross-curricular group, that we have found has the most widespread application is that of using *same/different* (i.e., using students' powers of discrimination to get them asking questions, or starting a discussion by presenting them with contrasting images or examples). In mathematics, a lesson starter I have used is to draw the images in Figure 4.7 on the board and ask the class to "Describe what you see." or "What is the same and what is different about these rectangles?"

A discussion has always ensued as to the fact that these rectangles have the same perimeter but different areas. Often a question such as, "What's the biggest area we could get with this perimeter?" has come naturally from a student and has provided the motivation for a number of lessons' work.

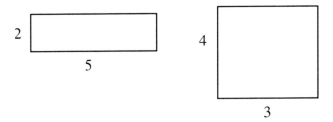

Figure 4.7.   Two rectangles —What's the same? What's different?

A teacher of History, Kay Attwood, described how her views about teaching year 7 classes had been influenced by the work of the cross-curricular group:

> I wanted to introduce this concept of *becoming a historian* in a wider forum and to promote an atmosphere where pupils take a far more active role in the classroom and it's less teacher led. ... trying to make sources more meaningful, that's sort of my own agenda. Our greatest problem is about helping pupils understand the questions surrounding the reliability of evidence at GCSE (national exams taken when students are aged 16). And obviously, my job is to start that on the first day in year 7... get the students to actively look at sources more and reach their own conclusions more. When I talk to them about issues they can see what I'm saying but then the minute I stop they don't think for themselves. And it's trying to get them to *think proactively*—Who wrote the source? Why was it written? When was it written? Is there bias? Is it subjective? Objective? And to use those terms to think about reliability.

One way she had used successfully of getting students to think proactively was starting a lesson by giving everyone two sources of evidence about the period they were studying. The first task was to comment about what was the same or different about these sources and the information they held. Students commented, among other things, on the distinction between one being a primary and the other a secondary source of evidence (without using those words) and a number of questions were raised that the students wanted to know about the historical period.

The cross-curricular group continues to meet. The format of meetings is always the same; a teacher gives a ten-minute lesson (using a beginning they would use with a class of students). The rest of the group engage (as themselves) in the activity and then discuss what teaching strategies they observed and examples of when/if they use strategies that are either markedly similar or different. Recently we became aware that almost all teachers of all subjects in our school want students to go through a cycle of activities broadly involving (a) questioning, (b) collecting evidence, (c) presenting evidence, (d) analyzing evidence, and (e) evaluating, which may lead back to questioning. We all used slightly different (sometimes contradictory) words and obviously different contexts but there seems to be a powerful commonality here that the group is going to try and exploit. As well as the TTA research leading to my involvement with teachers in other subjects in my school, Laurinda was interested to see if some of the findings from the TTA research about the creation of classroom culture could be used in working with a wider group of mathematics teachers.

LAURINDA: INTERLUDE 3

I led a successful bid to the Economic and Social Research Council (ESRC) to fund a project involving three teachers, three researchers and one teacher-researcher (Alf). The project's aims were:

1. to create year 7 mathematics classroom cultures which provoke a need for algebra;
2. to investigate the similarities and differences developed in each of the teacher's classrooms;
3. to investigate the nature and extent of the support needed from the collaborative group of teachers to plan their classroom activities starting from the students' powers of discrimination; and
4. to develop theories and methodologies to describe the complex process of teaching and learning.

Over one academic year, September 1999 to July 2000—which is split into three terms—we investigated the samenesses and differences in the developing algebraic activity in the classroom cultures of the four teachers by analyzing data from:

- collaborative group meetings of all seven once every half-term for a full day and corresponding through e-mail.
- collaborative group meetings of the four researchers, meeting for an evening session before the full-day meeting (pre-meeting) and for a half-day meeting after the full day meeting (post-meeting).
- videotapes of each teacher for one lesson in every half-term and researcher observations of teachers in the classroom at most once every two weeks in teacher/researcher pairs.
- interviews every half term of each teacher and six of each teachers' students in pairs, selected to give a range of achievement within the class.
- student writing while doing mathematics and at the end of an activity, about "What have I learnt?" Photocopies of all the "What have I learnt?" papers are collected from each teacher as well as all the written work of the six students interviewed.

Each researcher was responsible for viewing the data collected through one or more strands. The researchers looked for teaching strategies, student perspectives, algebraic activity, samenesses and differences in the classroom cultures, and teachers' use of same/different in planning to teach through students using their powers of discrimination. The overall structure was to support our looking at what students and teachers do in these classrooms. Schemes of work and organizational structures within the schools are different and it was not our intention to change them. The content of the lessons was decided by the teachers within those structures but during the day meetings there was time to plan together, given those constraints, to allow students to use their powers of discrimination. There were several methodological principles behind the design of the bid, which Alf gives his perspective on below in the context of his part in the research.

ALF: METHODOLOGY

As a researcher on the ESRC project, my strand was to look at teaching strategies. I was responsible for classroom observations of one of the other three teachers, and had access to all the video recordings, which all four researchers would often view together. One principle behind the design of the project was the belief that *we are what we do* (Varela, 1999). A practical consequence of this was that, when researching teaching strategies, I was not interested in teachers' planning, or what strategies they thought they were using, but only what I could observe from the video-taped lessons or lesson observations. In other words, I wanted evidence of what teachers did, rather than their beliefs or self-descriptions. This applied equally to looking at my own classroom.

The reason each researcher on the project took a different research strand was so that in our analyses we were able to take multiple views of a wide range of data:

> The aim here is not to come to some sort of "average" interpretation that somehow captures the common essence of disparate situations, but rather to see the sense in a range of occurrences, and the sphere of possibilities involved. (Reid, 1996, p. 207).

We often took one incident and analyzed it through our different strands and also told stories of the changes that were happening over time. This was achieved at the researcher pre-meetings by one researcher selecting a short extract from one or more video tapes that seemed significant to his/her strand. We would talk through the details of what we saw, often spending a long time agreeing on the text of what was said before moving to a discussion of our different analyses. These joint viewings of data were a time when I could check out, for example, whether anyone else could observe teaching strategies using the definition I was developing. Did the other researchers see the same or different? We would view the same video extracts in the subsequent group meetings of seven teachers and researchers.

When watching video segments for the first time I was surprised consistently by how inaccurate my initial impressions were of what students/ teachers had said on these video tapes. I became much more wary, as a teacher, of assuming I had heard and understood what one of my students said in the classroom.

As researchers, through our detailed study of the data, we noticed many overlapping and interconnected patterns and developed theories from these observations. It seems of vital importance that the purpose of these theories was to inform our future observations and the actions of all the teachers on the project. We would continue talking about and developing a particular theory for as long as it remained useful for these purposes.

> ... theories and models ... are not models of ... they do not purport to be representations of an existing reality. Rather they are theories for; they have a purpose, clarifying our understanding of the learning of mathematics for example, and it is their usefulness in terms of that purpose which determines their value. (Reid, 1996, p. 208)

There was no sense of there being a "best" theory for this work. I recognize what is useful for my teaching by noticing what I am doing in my classroom. What is not useful simply does not happen.

ALF: FINDINGS OF THE ESRC PROJECT

There is not space to go in to much detail about the findings of the ESRC project. There was evidence that all four of the project's aims were met. In all four classrooms, we observed students finding a need for algebra, which was a key objective. The four factors from the TTA study were recognized by all the teachers as being important in achieving this, particularly the central role of establishing a level of metacommunication in the classroom in which purposes could be expressed, both for the year (e.g., becoming a mathematician) and for individual lessons or sequences of lessons (e.g., "Given any rule, can you predict where the graph will go?"). One of the teachers described things they would take away from the project:

> ... the widening of the meaning of algebra as a language for being a mathematician and thinking mathematically. This is linked to the mathematical behaviors such as systematic approaches and using

notation as a shorthand. These are algebraic but not the traditional use of letters, not just manipulating symbols.

Evidence, such as this statement, of personal change and development is ultimately what the project set out to provoke.

LAURINDA: INTERLUDE 4

What do I learn through working with Alf and other teachers co-teaching and co-researching in their classrooms? In the final report to the ESRC for the project reported here I wrote:

> The results of this project ... are found in the voices of the teachers and the students, in the patterns of communication from the interviews, observations, video tapes and discussions at the day meetings. These patterns of communication are closely linked to their actions, as students doing mathematics or as labels for teaching strategies. For Bateson (2000), "learning is a communicational phenomenon" (p. 279).

When I wrote this, I was arguing that the results of the project were in the learning of the participants either doing mathematics or learning about teaching mathematics. The process is what interests me, that of teachers finding ways of supporting the learning of mathematics of their students. For any individual this journey will be different. For me, learning is a process and as I spend time in classrooms looking at what students and teachers do, over time I come to see more and am able to make ever more finer distinctions from more developed awarenesses. These awarenesses support me directly in my job as a teacher of people learning to be mathematics teachers, but that is not the sole motivation. I am interested in how people learn mathematics and continue actively to learn myself. I learn best in communities of people interested in what I am interested in, where the communication is open and creative. I would not teach as I do now without seeing possibilities for action in Alf's and others' classrooms that I could not ever have conceived of from my own individual history. I notice difference and the consequences of those different actions and realize that difference does not have to be judged as "bad" but can allow me to grow and learn.

ALF: CONCLUSION

All the research I have been involved in has begun with identifying aspects of my teaching that I want to investigate. Research has always been a process of learning about teaching. At the heart of this learning is the discipline of trying to separate my judgments about what I see from detailed description and, since this is never ultimately possible, trying to become aware of what I am bringing to my observations—where I am coming from that means I notice what I do. In working with others on developing teaching I try to impose the same discipline. I see the teaching/researching process as a questioning of assumptions in an attempt to see more detail in the classrooms (my own and other's) that I observe, hence the power of working with others—having common experiences from which to share observations.

Laurinda and I have frequently written up our work in proceedings of the British Society for Research into Learning Mathematics (BSRLM) and recently there has been a critical review of BSRLM papers (Nickson, 2003). We leave the last word to this review:

[Brown and Coles'] concern is with the development of a "metacommunity" within a classroom where pupils are engaged in doing algebra. ... The research initially draws on the work of Kieran (1996) and although the project focuses strongly on pupils learning algebra, a major thrust of the work is to help teachers to reflect upon their interactions with their pupils and to develop theories as a result of this analysis. The aim is not to lose sight of the detail of pupil actions and their algebraic development in the totality of the classroom situation. A characteristic of the study is that theories are developed throughout in response to the data that is collected as it progresses. (p. 33)

... worldwide research projects in the development of teaching in mathematics education tend to encourage models of critically reflective practice leading to the development of communities of enquiry together with critical intelligence in them. This type of research is well illustrated by the work of Coles and Brown (e.g., 1997, 1998). [Their] initial paper (1998) relating to their ongoing study includes a reflection on what it is like for teachers and researchers to work together. In the long term, this project is concerned with the development of a culture of algebraic thought and activity within the classroom; however, the reflections on the initial stage on the role of teacher as researcher indicate the fruitfulness of such a partnership and the results of critical reflection (p. 63).

As well as positive outcomes in terms of classroom learning, the study in its entirety is a very good example of the benefits of collaboration over time between a teacher and a colleague for whom research is part of his or her professional life. The fact that the BSRLM community as a whole gains from it is an added bonus to the profession as a whole, both in terms of development of individuals concerned and in terms of the building up of a working understanding and partnership between these two areas of mathematics education. (p. 64)

ACKNOWLEDGMENTS

Our thanks goes to the head teacher and governors of Kingsfield School who have consistently supported and encouraged this research, and to all the students and teachers who have been involved.

External funding has been provided by: (a) the Economic and Social Research Council for *Developing Algebraic Activity in a Community of Inquirers* (R000223044), awarded Laurinda Brown, Rosamund Sutherland, Jan Winter, Alf Coles, and (b) the UK's Teacher Training Agency for *Developing a Need for Algebra*, awarded to Alf Coles.

REFERENCES

Arcavi, A. (1994). Symbol sense: Informal sense making in formal mathematics. *For the Learning of Mathematics, 14*(3), 24-35.

Bateson, G. (2000). *Steps to an ecology of mind.* Chicago: University of Chicago Press.

Brown, L., & Coles, A. (1997). Interpretations of a classroom vignette or what does reading about someone else's theories-in-action do for you. *Proceedings of the British Society for Research into Learning Mathematics, 17*(1 & 2), 23-29.

Brown, L., & Coles, A. (2000) Complex decision-making in the classroom: The teacher as an intuitive practitioner. In T. Atkinson & G. Claxton (Eds.), *The intuitive practitioner: On the value of not always knowing what one is doing* (pp. 165-181). Buckingham: Open University Press.

Coles, A. (2001). Listening: A case study of teacher change. *Proceedings of the British Society for Research into Learning Mathematics, 21*(1), 1-6.

Coles, A., & Brown, L. (1998). Developing algebra: A case study of the first lessons from the beginning of year 7. *Proceedings of the British Society for Research into Learning Mathematics, 18*(3), 17-22.

Coles, A. & Brown, L. (2000). Needing to use algebra. In C. Morgan & K. Jones (Eds.), *Research in Mathematics Education Volume 3: Papers of the British Society for Research into Learning Mathematics.* London: British Society for Research into Learning Mathematics.

Holt, J. (1964). *How children fail.* London: Penguin.

Mason, J. (1996) Personal enquiry: Moving from concern towards research (Monograph R). *ME822 Researching Mathematics Classrooms,* Milton Keynes: The Open University.

McIntosh, A., & Quadling, D. Arithmogons. (1975) *Mathematics Teaching,* 18 - 23. Association of Teachers of Mathematics, March 1975.

Nickson, M. (2003). *A review of British Society for Research into Learning Mathematics research, 1995-2002.* Totton, Hampshire: British Society for Research into Learning Mathematics.

Reid, D. A. (1996). Enactivism as a methodology. In L. Puig & A. Gutiérrez (Eds.), *Proceedings of the Twentieth Annual Conference of the International Group for the Psychology of Mathematics Education* (4th ed.). (pp. 203-209). Valencia, Universitat de València:PME.

Royal Society and Joint Mathematical Council (RS/JMC). (1997). *Teaching and learning algebra pre-19.* Report of a Royal Society/Joint Mathematical Council Working Group, London: Royal Society, Carlton House Terrace.

Varela, F. (1999). *Ethical know-how.* Stanford: Stanford University Press.

CHAPTER 5

A DEVELOPING MATHEMATICAL COMMUNITY

Vicki Walker
Indiana University-Purdue University at Indianapolis, Indianapolis, IN

with

Beatriz D'Ambrosio
Miami University, Oxford, OH

and

Signe Kastberg[1]
Indiana University-Purdue University at Indianapolis, Indianapolis, IN

As I attempt to piece together an account of this project, a story swells up within me and summons me to release it. It is a story I find overwhelmingly complex and burdensome to unravel. As I try to capture the essence of this story, I take numerous backward glances into my mind's view of my own classroom setting, in an effort to snag a critical moment or significant

Teachers Engaged in Research
Inquiry Into Mathematics Classrooms, Grades 6–8, pages 81–98
Copyright © 2006 by Information Age Publishing

81

detail worthy of being transcribed. What I see instead are the faces of students, the students with whom we worked throughout this project. This brings to the foreground what is most important, and that which is central to the writing of this paper.

The story to be told here is about the students. It is about individual students whose mathematical thinking became the driving force for the instructional decisions made. It is a story about mathematical empowerment, not just for a few, but in particular for the non-compliant, underachieving students in our population. It is a story about the evolution of a mathematical community, a community in which rich conversation and justification of ideas was the rule of thumb, rather than the exception to the rule.

The backdrop for this story is an urban K–8 school in a large Midwestern city in the United States. The middle school grades had recently been added to an already existing K–5 school, making the school's identity as a K–8 option within the school system not only new, but rather unique as well. The school's population numbered about 500 primarily African American students. The particular focus of this project involved the school's sixth, seventh and eighth grade students who were grouped in four multiage classes of about 25 students per class. Students were not grouped by age, grade or ability, and inclusion students were part of each class make up.

I was the mathematics teacher-researcher for these middle school students and Beatriz D'Ambrosio was the collaborating researcher who attended the classes twice a week during the entire school year. It is important to point out that Beatriz and I had actually begun a collaborative project during the previous school year when I had a different teaching assignment within the same public school system with a group of seventh grade students. We had launched this initial collaborative project after having numerous conversations and raising a plethora of questions about the choice and implementation of tasks for mathematical explorations with middle school students. We had discussed what constituted big ideas, what were our goals for students' learning, what were the best strategies we knew for achieving those goals, how we would engage students in classroom activities, and many more issues that seemed related to our plans for enhancing student learning. The overall goal of our work together had been to understand better the teaching and learning of mathematics in middle schools.

The initial collaborative project in the seventh grade classroom helped establish the framework for our continued, more formalized collaboration in the multiage middle school setting. Beatriz' regular classroom visits, her observations and copious field notes pertaining to the students and their work, our ongoing instructional planning and debriefing sessions, and the raising of a new myriad of questions about student learning embodied our research endeavors and served to shape our project.

DEVELOPMENT OF A COMMUNITY OF MATHEMATICIANS

My account of this project centers itself primarily around three different developmental phases, though no clear delineation exists between these phases. In retrospect, there seems to be a general trend from a rather chaotic frenzy within the classroom to a more cohesive flow of events. Yet there is no fluid line of demarcation that can be used to discern key moments of transition. As a means of unraveling some of the complex dimensions of this story about the students, it is perhaps helpful to zoom in on the various phases of this project.

The Preparation Phase

A measure of intense planning and *preparation* took place as the school year began and serves to characterize the initial phase of our project. Conversations with Beatriz increased, both in number and intensity and we began to formulate ways in which a collaborative effort might take place in this new setting. Yet, my focus during this period of time seemed to be primarily on the logistical details of getting the year started and getting my room set up for instruction. There were so many decisions that needed to be made, most of which were made initially devoid of any real knowledge and understanding of who the students were. Even seemingly insignificant decisions were paramount in my mind as I set out to organize and structure the physical learning space and the intended instruction.

Decisions about the classroom itself involved the arrangement of desks in groups of four, positioning them around the overhead projector and display area. In this way, group work and the sharing of ideas to the whole class could take place. I set up and utilized shelf space so that notebooks for student journaling, files for student portfolio collections, and various mathematical manipulatives could be both visible and accessible.

Decisions about essential classroom materials served to bridge my thinking between the mere physical structure of the classroom to the more important decisions about curriculum and instruction. I was fortunate enough not to be held by the school to adhere to a specific curriculum series and was given full autonomy in this regard. However, it was the sustained communication and level of involvement with Beatriz throughout this period of time that greatly facilitated my decision-making process. Decisions about instructional materials, specific tasks and the sequencing of topics came about as Beatriz and I considered and discussed the major issues related to the school setting, namely that it was a newly formed middle school, that classes would be rather small in size, ranging from 20–25 students, and that classes would be multiage, with a mix of sixth, seventh, and eighth grade students. We were committed to the notion of building a mathematical community in which all students could be involved in the same exploration regardless of age, grade or ability. We were especially committed to the goal of *not* separating students by any such factors. Instead, our primary goal was to facilitate mathematical explorations in which the students' differing levels of ability, experiences and interests would broaden, deepen and overall enrich their mathematical conversations. Thus, we decided on a three-year revolving curriculum with the following content focus: Year 1: Algebraic Thinking, Year 2: Geometry and Spatial Reasoning, and Year 3: Data, Statistics and Probability. This revolving curricular design would allow all students to engage in the same three-year cycle without regard to a specific age, grade or ability level.

Once the overall structure and plan of instruction had been formulated, emphasis shifted more heavily to the tasks themselves. Beatriz and I were curious about the kinds of tasks that might lead to high levels of engagement by the students, and in turn result in more sense making for them about important mathematical ideas. We had established some characteristics of the kinds of tasks we deemed to be worthwhile. We wanted tasks to involve problem solving and be open ended in nature, conducive to group work, engaging and challenging, and centered on the big ideas we had already outlined. Further, we wanted to be sure tasks would encourage reflection and communication, possess the potential for *residue*, and have an entry point for all students (Hiebert et al., 1997).

As this phase of planning and preparation continued, a number of obstacles occurred that began to clutter the path and make for less than smooth sailing into the first few days and weeks of school. One of the biggest dilemmas was the fact that no schedule yet existed that would accommodate the 80-minute block periods, the specials (e.g., art, music, physical education) that needed to be squeezed in, and the elementary school's schedule all in the same structure at the same time. The fact that middle school classes were going to be multiage complicated the scheduling issues. So while we were planning for 80 minute periods, we were unsure how they would be managed, and whether or not we would only be able to meet groups of students on a rotating basis rather than every day. Additionally, physical space allocated to the middle school was minimal and there were few structures in place reflective of a middle school concept. I recall feelings of bewilderment and uneasiness before the first day had even begun.

Regardless of my lack of readiness, the first day came and quickly smeared into the first few weeks of school. One tentative schedule after another was put into place until finally, after a few weeks of school had passed, the schedule morphed into a workable situation. We ended up with 80-minute sessions four days a week and one 40-minute session per week for each of the four groups of students.

The Planting Phase

As the school year began and the names on my class lists became real students to me, I became consumed with an entirely new set of issues and concerns. This gave way to a new phase of development in terms of our project. It was a critical time for new beginnings, for tiny seeds to be planted which would later become incredible landmarks of growth and development. This particular phase was a *planting* phase in several ways. I found myself working to establish routines, working to establish relationships with co-workers and students, and working to establish certain notions about mathematics class. The cultivation of these routines, relationships, and notions was many times painstaking and exhausting.

As Beatriz' regular visits to the classroom became recognized by the students and the school community at large, relationships began to sprout across many different levels. The students sought her out in order to share bits and pieces of their lives with another listening adult, administrators and other faculty became inquisitive about our project and time together, and the interaction between Beatriz and I deepened as a genuine fascination emerged for both of us with regard to the students, their thinking, and the development of their mathematical ideas. We soon were in constant dialogue about what the students had done in class. We became driven by conversations about what *we* thought *they* thought.

Admittedly, my focus during this phase was still primarily on student management, engagement and evaluation. I spent a great deal of time addressing issues related to personality, social climate, and adolescence, in general. Such concerns often took precedence over regard for mathematical decisions and students' mathematical thinking. I struggled to stabilize the classroom environment and in general, felt pulled in all directions. I struggled to find a balance between being flexible and being overextended like a rubber band ready to snap.

There were frequent team meetings prior to the start of the school day, additional classes to be covered for teachers who were absent, weekly school convocations, lunch and recess duties, homeroom students to manage, students assigned to me as advisees, special interest clubs to organize and manage, and a wealth of other issues that needed to be assimilated into the regular flow of the teaching day. I typically began each day feeling as if I was stepping onto a downward escalator in an attempt to climb upward. I was constantly moving and exerting energy, but was plagued by some mysterious force that tugged at me and pulled me in an opposing direction. I felt that if I let up in my efforts, or paused long enough to catch my breath, I would slide backward beyond the point of recovery.

Interestingly, as I became more pulled apart and frazzled as a member of the school community, I simultaneously became more grounded and focused on the individuals with whom I spent time pondering mathematics. Tiny seeds had given way to growth. The students began to respond in a positive way to the warm-up problems which were regularly posted on the overhead as they walked into the classroom. They began working in their groups without coaching on my part to do so. The students grew accustomed to Beatriz' presence and her active participation in their learning experiences. They began to look for her on a daily basis and to question if it would be a day in which she would be visiting. On days when Beatriz visited, they began to take notice of the ways in which Beatriz and I interacted with each other. They grew comfortable with the way in which Beatriz and I circulated around the room and listened as they worked on various tasks.

I began to sense that when students came into the classroom and the door closed, there was a feeling of belongingness and responsibility to one another. We were temporarily exempt from the murkiness of the overall *school climate* and comforted by the familiarity of our own developing *classroom climate*. The seeds had taken root. These very tiny, seemingly insignificant seeds related to the development of classroom routine and expectations. They also related to the building of trust and of relationships. They were seeds that produced viable, tangible results that, in turn, allowed for new layers of our project's focus to be peeled back for scrutiny.

The Shifting Phase

As issues related to the classroom routine and expectations began to fade into our peripheral view, new ones appeared before us as if in the spotlight of our minds. A sense of visual acuity seemed to illuminate the students to us now in new and exhilarating ways. The students argued with one another over their solutions to mathematical problems, rather than over petty issues such as pencil stealing and name calling. They asked to share at the overhead rather than waiting to be called on. During such sharing times, they responded to ideas their classmates shared by raising questions and debating issues rather than shrugging their shoulders and merely biding their time.

These subtle but monumental changes in the overall flow of classroom events characterize yet another phase of development in this collaborative project, which I describe as the *shifting* phase. Shifting occurred at all levels, in varying magnitudes, and with multiple outcomes. There were shifts in the focus of our project away from the tasks and toward the nature of discourse that was taking place. There were shifts in our own thinking about the complexities and connections between mathematical ideas. There were shifts in the ways we thought about and listened to the students' mathematical thinking. Finally, there were shifts in the students themselves, in who they were as learners, as individuals, and as members of the mathematical community.

As part of all this shifting that occurred there were two particular revealings we found the most noteworthy and impacting. The first revealing came from a shift within the community when students began to listen to one another differently. Student conversations both fueled and directed the sense-making process. Students listened to one another in a sense-making way, and then used each other's thinking in order to make sense of mathematical ideas for themselves. The second revealing came about as a gradual shift in leadership occurred among the students. The students who emerged as leaders within the mathematical community were not the same leaders within the larger school community. Instead, the leaders of this sense-making, mathematical community were typically non-compliant, underachieving students.

The data used to ground this project consists of field notes from both Beatriz and me, which included detailed transcripts of students' conversations and presentations, and their written work. A closer look at such students' work will serve to highlight this, and will further illuminate the impact of student discourse on student understanding.

REVEALINGS OF A COMMUNITY OF MATHEMATICIANS

In this section we describe the mathematical thinking of a collection of students considered by the larger school community to be non-compliant, underachieving students. In particular we describe the students' solutions to a single problem presented late in the school year (May 1, 2001).

Three pirates were hunting for treasures and found a hefty treasure. They decided to split it evenly. Since it was late they thought they should sleep and split the treasure the next morning. During the night one of the pirates got up and took ⅓ of the treasure and ran away. A second pirate woke up and took ⅓ of what he saw and ran away. The third pirate got up in the morning and looked for the others. Trusting that they had each taken their fair share, he took what was left for himself. Did all three pirates get their fair shares? If not, which pirate got more and which got less?

Following some time to think and discuss their work in their small groups, the students were eager to present their findings to the class. The presentation and subsequent discussion took place over two class periods with some of the shared work summarized by the students in their journals during the second day of work.

Equal Shares

As one might expect, several of the students developed an equal shares perspective on the pirate problem. Figure 1 provides a view of the typical first solution for students whose thinking hovered around equal shares.

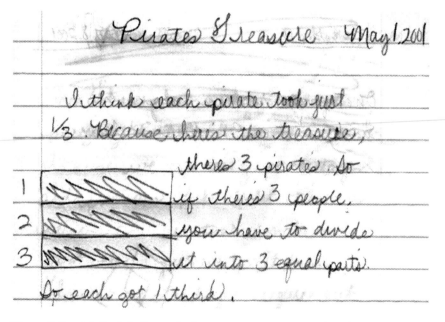

Figure 5.1.   Melinda's first solution response; an example of several students' initial individual thinking

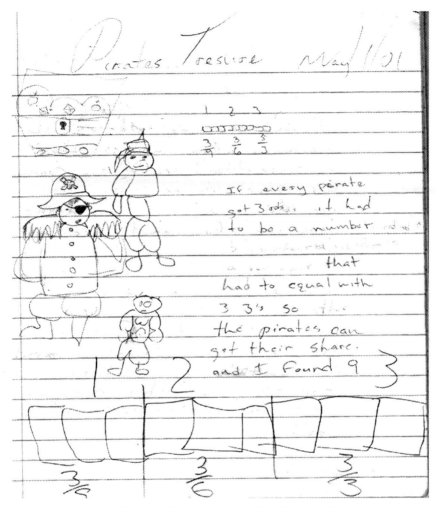

Figure 5.2.   Aaron's journal entry, upon which he based his presentation

One interpretation of Melinda's solution is that she saw the treasure as an ever present whole. Thus, each pirate was aware of how much treasure he or she left and only took that portion. However, not all students that reasoned the pirates got equal shares represented the portions as part of a single whole.

For some students, like Aaron whose work is presented in Figure 5.2, identifying a concrete whole was the first stage in their numeric reasoning.

Aaron identified nine as a number with three threes in it. This allowed him to allocate three parts of the treasure for each pirate. He named each of these collections of three pieces with a fraction based on the remaining

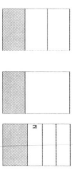

Figure 5.3.   A replica of Miguel's work at the overhead projector.

whole. Thus, the first pirate took three of the nine pieces of treasure or 3/9. The second pirate, seeing six pieces, took three of them. The remaining pirate took what was left of three of three. Aaron concluded the shares were equal because each pirate got three pieces.

Unequal Shares

Not all students agreed that the pirates shared the treasure fairly. Miguel was the first to suggest this alternative interpretation of the problem. For Miguel, each pirate faced a different whole. A replica of his work at the overhead projector appears in Figure 5.3.

Miguel explained that after the first pirate took ⅓ of the treasure, the second pirate only took ⅓ of what he saw, identified in Figure 5.3 by strip labeled with a two (representing the second pirate's share). Miguel said, "This is less than ⅓. So the third pirate got the most." When Miguel was asked how he knew that, he referred to the picture as his proof and said, "Well look at the picture. It's pretty close to one-half."

Although some students agreed with Miguel's reasoning, Matt wanted to present what he viewed as a different answer. Using the notes he had made in his notebook (see Figure 5.4), Matt suggested that the treasure be divided into sixths. However, as he began explaining his work he realized that sixths would not work and he proceeded to add another row of three pieces to his already drawn sixths (see Figure 5.5).

This change in his diagram confused some of the other students despite his explanation that the first pirate was to get three pieces, the second two, and the final pirate would get four parts of the treasure. Aaron and others argued that 2/9 was not ⅓ of six. Matt's initial drawing and focus on sixths seemed to prevent others from following his thinking. He grew frustrated

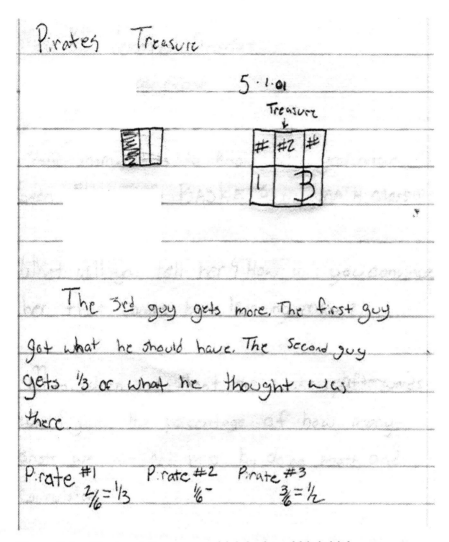

Pirates Treasure

5 · 1 · 01

Treasure

The 3rd guy gets more. The first guy got what he should have. The second guy gets 1/3 of what he thought was there.

Pirate #1   Pirate #2   Pirate #3
2/6 = 1/3   1/6   3/6 = 1/2

Figure 5.4.   Matt's solution upon which he based his initial presentation changing his solution in mid-sentence

Figure 5.5.   A replica of Matt's work at the overhead projector.

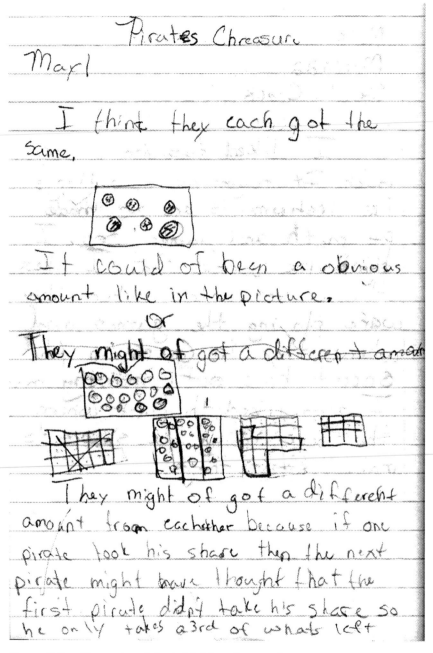

Pirates Chreasure

Marl

I think they each got the same.

It could of been a obvious amount like in the picture.

or

They might of got a different amam

They might of got a different amount from eachother because if one pirate took his share then the next pirate might have thought that the first pirate didn't take his share so he only takes a 3rd of whats left

Figure 5.6.   Sharon's original thinking that set the stage
              for her receptivity to Matt's explanation

Figure 5.7.   A replica of Sharon's work at the overhead projector

that others could not follow his thinking, yet all the more convinced of his own solution.

Sharon presented after Matt with only moments left in the class period. Sharon, a student whose initial solution was based on equal shares, had revised her work in light of other students' presentations. Her evolving thinking was reflected in her notebook (see Figure 5.6).

In her presentation, Sharon generated a rectangle containing nine equal parts. She then shaded three of these (see Figure 5.7) to represent the first pirate's share.

She explained "Then you can see that the second pirate takes ⅓ of what he sees or ⅓ of the part that's not shared. (She pointed to two unshaded sections in the first column.) And that's 2/9."

This understanding came from Sharon's careful listening to all the presentations and the questions that the students were asking of each other. This was not her original solution, but it evolved during the presentations.

The students struggled to make sense of the solutions presented by their peers and in so doing, became conversationally engaged with one another. The collective conversation about this problem and the fact that the students were listening carefully to each other's presentations was evident. The interactions were a critical component of the students' sensemaking process.

The influence of the students' conversation is evident in their final thoughts recorded in their journals. Jeremy originally reasoned that the pirates took fair shares (see Figure 5.8), but on Day 2 he identified the third pirate as getting the most.

This is one example of student work that was influenced by Aaron's solution. Although Aaron reasoned that the pirates each took fair shares, he appeared to identify a new whole following each share's extraction. Of the nine original parts of the treasure 3 constitutes ⅓. Three of six parts is ½ of the remaining treasure and 3 of the remaining 3 is all of the treasure. Jeremy took Aaron's solution (which matched his original thinking) a step further. In light of the other explanations, he realized that the shares were of different sizes. His notation of 2/6 for the second pirate's share, in fact represents 2/6 of 6/9 of the original treasure. His notation of 4/4 for the

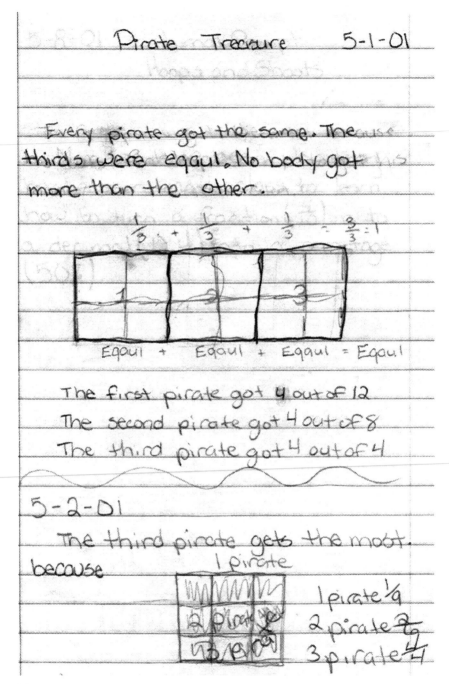

Figure 5.8.   Jeremy's initial solution and his thinking after
              having listened to the presentations

third pirate's share, in fact represents 4/4 of 4/9 of the original treasure. This thinking is quite sophisticated and although Jeremy did not have the symbolic language to represent his thinking in his journal, we were able to see how much growth he underwent as a result of this experience.

CONCLUSION

The revealings of our mathematical community are many, the most significant of which is the rich mathematical thinking shared by the students. In fraction problems with changing wholes the default position is often to make sense of the situation using reasoning based on a consistent whole. Because this mathematical community had shifted to one that focused on the viability of each member's thinking, sharing solutions became an opportunity for each member to develop his or her mathematics. In this case, students who initially reasoned using a single whole were able to hear peers' solutions and use the results of that hearing to identify changing wholes and represent parts of the wholes.

Researchers have noted that use of mathematical concepts in novel situations is evidence of understanding (Brown, Bransford, Campione, 1983). In this short excerpt of student mathematicians generating, discussing, and adapting their own thinking, we see evidence of developing understanding. In addition, this community uses discussion to build arguments not unlike the communities of mathematicians described by Thurston (1998). In both communities, problems remain unsolved until the mathematicians solve them. Solutions are generated first for one's self and then, thinking is represented for others to read. Finally, presentations are made and the merits of various solutions are discussed. Through this process mathematicians—be they middle school students or university professors—develop their ideas about the mathematical objects they use or develop to make sense of the problems they attempt to solve.

The episode involving the "Pirate's Treasure" problem is just one example of several we witnessed in which the students who were the most instrumental in impacting the sense-making process of others were students who typically did not have this kind of leadership role in the larger school community. Yet, within this mathematical community they demonstrated a great deal of exuberance and confidence when presenting their ideas. Their classmates were not only receptive to them as presenters, but were also attentive to and invested in the thinking presented. The overall conversation that ensued was fluid and authentic.

The emergence of new leaders within the classroom community gave way to richer, deeper levels of mathematical discourse. The level of trust that evolved as a result of this discourse yielded greater involvement on the part of students who typically did not or would not contribute during class sessions.

We had begun this collaborative project with the overall goal of better understanding the teaching and learning of mathematics in middle schools. This classroom of multiage, middle school students gave us a much deeper understanding of what it means to be an effective, powerful mathematical community.

We had begun the project with a concentrated focus on the nature of classroom tasks. These students' interactions with one another pulled our attention away from specific concern about the tasks and captured our full attention toward a new focus on the nature of mathematical discourse.

Finally, we had initially formulated our ideas about this project without any knowledge of the students or who they are as mathematical thinkers. Students like Aaron, Miguel, Matt and Sharon have taught us much about the notions of equity within the classroom and have shown with total clarity what it means to become mathematically empowered.

FINAL REFLECTION

Looking back, I realize I gained a much deeper understanding about the nature of teaching primarily because the emphasis of what Beatriz and I explored together was not on the teaching but on the students. It was the collaboration with Beatriz that pushed me to focus and to reflect on what was happening with regard to the students' mathematical thinking. It was the students' thinking, their responses and their interactions with one another that opened up for me new realms of thinking and operating as a teacher. The students' mathematical ideas and understandings challenged me to make sense of the mathematics in deeper, more substantial ways. Beatriz and I had innumerable conversations both about the mathematical ideas and the students' understanding of those ideas. As a result of these conversations with Beatriz, I became more and more intrigued with the students' mathematical thinking and more attuned to the art of *listening to* the students as opposed to the practice of *talking at* them. (See D'Ambrosio, 2004 for a more extensive discussion of the practice of listening.) Interestingly, we watched the students become authentically engaged with one another and with the mathematics at hand. The particular mathematical tasks given to them did not seem to matter. What mattered was the student discourse and their willingness to grapple with mathematical ideas.

I find it difficult to reflect on this experience and this research project without thinking that I learned far more from it than did any of the students. There clearly was evidence of tremendous growth and learning for the students. Yet, what I feel I learned from the students and from working closely with Beatriz is completely inexpressible. It is difficult to imagine setting foot now in a classroom without it being a collaborative endeavor. The countless hours spent talking about and thinking about the students and their developing mathematical ideas would never have taken place if not for our collaborative project. The deepening of my own mathematical understandings would have been nonexistent. I am still awakening to mathematical notions and to new understandings about the teaching/learning process as a direct result of this project. Clearly, our research did not take the shape of a specific question we sought to answer. Instead, our research was a process, a journey, an evolution of numerous ideas and questions, ever changing, ever growing, but always guided by the students and by what *we* learned from *them*.

NOTE

1. This chapter is the story of a middle-school mathematics teacher-researcher, Vicki Walker, who collaborated with colleagues from a local university, Beatriz D'Ambrosio and Signe Kastberg, to make sense of the mathematics learning of her middle-school students. Vicki tells the story of her involvement in this collaborative endeavor, which occurred in two different instances. First, she and Beatriz collaborated in the classroom setting, planning instruction, interacting with the children, debriefing the teaching episodes, collecting and analyzing data. Then Signe joined the collaborative team for the data analysis process and offered another lens through which the data would be analyzed. Hence, the shift in the story from first person to third person, as Vicki tells her story and the team presents its findings.

REFERENCES

Brown, A., Bransford, J. D., Ferrara, R., & Campione, J. C. (1983). Learning, remembering, and understanding. In J. Flavell & E. Markman (Eds.), *Handbook of child psychology* (Vol. 3, pp. 77–66). New York: Wiley.

D'Ambrosio, B. S. (2004). Preparing teachers to teach mathematics within a constructivist framework: The importance of listening to children. In T. Watanabe & D. R. Thompson (Eds.), *The work of mathematics teacher educators: Exchanging ideas for effective practice* (Vol. 1, pp. 135–150). Monograph Series. San Diego, CA: Association of Mathematics Teacher Educators.

Heibert, J., Carpenter, T. P., Fennema, E., Fuson, K. C., Wearne, D., Murray, H., et al. (1997). *Making sense: Teaching and learning mathematics with understanding.* Portsmouth, NH: Heinemann.

Thurston, W. P. (1988). On proof and progress in mathematics. In T. Tymoczko (Ed.), *New directions in the philosophy of mathematics: An anthology* (pp. 337–355). Princeton, NJ: Princeton University Press.

CHAPTER 6

DRIVING WHILE
BLACK OR BROWN

The Mathematics of Racial Profiling

Eric Gutstein
University of Illinois-Chicago, Chicago, IL

In the dramatic film version of the life of Rubin "Hurricane" Carter, Denzel Washington plays Hurricane, the great middleweight boxer from Paterson, New Jersey who spent 20+ years in prison for a crime he did not commit. The film portrays the following story: In 1966, two gunmen invade a Paterson bar and shoot fatally three people. At about the same time, Hurricane and a young friend, John Artis, are leaving a party not too far away. As they drive home, the police pull them over. The officers shine their light inside the car and recognize Hurricane, a local hero. One of them says quickly, "Oh sorry, Mr. Carter, we didn't know it was you. We were looking for a couple of Negroes in a white car." To which Denzel dryly responds, "Any two will do?"

Teachers Engaged in Research
Inquiry Into Mathematics Classrooms, Grades 6–8, pages 99–118
Copyright © 2006 by Information Age Publishing
All rights of reproduction in any form reserved.

RACIAL PROFILING AND MATHEMATICS CLASS

Racial profiling. "Driving While Black (or Brown)," as it is sarcastically called. According to the American Civil Liberties Union (ACLU), racial profiling is "the practice of using race, ethnicity, national origin, or religion as the primary factor in deciding who to subject to law enforcement investigations" (ACLU, 2003). Black men, in particular, know the refrain all too well, and Denzel's words ring true for many African Americans and Latinos. Throughout the United States, various civil rights organizations have documented thousands of cases of *discretionary*[1] police stops of people of color. The ACLU even has a form on its web site for people to file racial profiling complaints. In addition, they have filed suit in six states and, along with other civil rights organizations, at the time of this writing were urging the public to support the federal *End Racial Profiling Act of 2004*. Racial profiling is a serious problem that impacts communities of color across the United States and, in a post September 11 world, has affected Muslims, Arabs, and many others of Middle Eastern or South Asian descent who might be taken as such.

But what does racial profiling have to do with mathematics class? Or with the National Council of Teachers of Mathematics *Standards*? In fact, quite a bit. Racial profiling is not only a political phenomenon, it is also a mathematical concept that cannot be fully understood without a mathematical analysis. It relies on the mathematical idea of *expected value* and can be explained as follows: if one is examining police stops of vehicle drivers, then given the percentage of a particular racial group within a driving population, one would *expect* that the number of discretionary stops would reflect that percentage, within an appropriate range. The mathematics of probability theory tells us this. Thus, if 60% of the drivers on a particular road were African American, one would expect (and hope) that, over time, roughly 60% of the stops would be of African Americans. In the situations where the ACLU has filed suit, however, the data are dramatically different. For example:

> [I]n Illinois State Police District 13, which covers seven counties southeast of St. Louis, Hispanics comprise less than one percent of the local driving-age population, yet they represent 29 percent of all people stopped by these officers for speeding less than five miles above the speed limit. (ACLU, 1999)

Even a rudimentary mathematical analysis would suggest the virtual impossibility of these numbers occurring randomly.

Furthermore, consider a different scenario. If 30% of the drivers in an area were African American, but of 10 police stops, only 1 person was African

American, one could not easily conclude that racial profiling did *not* exist. However, to understand that one would need some understanding of the law of large numbers. Additionally, students would need a relatively good understanding of probability and statistics to design and run a simulation that replicates random police stops of those 30% African American drivers. Students investigating the mathematics of racial profiling would encounter a number of important statistical concepts including the law of large numbers, expected value, theoretical and empirical probability, and probability simulations—all important ideas as outlined in the NCTM (2000) Grades 6–8 *Data Analysis and Probability* Standard.

Thus, racial profiling is a rich area for sophisticated middle school mathematics. However, it is not *just* a mathematical topic—nor is it one of the types of problems suggested in the *Data Analysis and Probability* Standard (NCTM, 2000). The NCTM suggested that students experiment with, and collect data on, how far a paper airplane with one paper clip flies, as compared to one with two paper clips. But the NCTM's (2000) *Connections* Standard also recommended, "School mathematics experiences at all levels should include opportunities to learn about mathematics by working on problems arising in contexts outside of mathematics" (p. 65). Racial profiling is such a context, but it goes beyond the "real-world" mathematics called for in the *Standards*, because it also allows students to study and learn about the social ramifications of those outside contexts. It is an investigation into a real-world issue that is ultimately political in nature, that asks students to think about issues of racism, injustice, police conduct, and more. I say more about this below, but first I describe how I taught a project on racial profiling and studied the process in my own classroom.

TEACHING MATHEMATICS FOR SOCIAL JUSTICE AT RIVERA SCHOOL

As part of my work as a university-based mathematics educator, I have periodically taught a mathematics class in a Chicago public school, Rivera (a pseudonym), located in a Mexican immigrant community, and I studied my own teaching and my students' learning. I taught at Rivera for a total of approximately four years over the six-year period from 1997 until 2003, either seventh or eighth grade. While the classroom teacher, I had the full responsibility for everything associated with my class—instruction, planning, assessment, report cards, parent contact, standardized test preparation, etc.

I originally started working with teachers at Rivera in 1994 through professional development activities with the school's mathematics teachers. As I worked with them, it became apparent to me that some teachers had close ties with students and the community and seemed able to build on students' cultural knowledge and experience. After discussions with teachers, other faculty members, and community residents, several of us initiated a collaborative research project in which we used the theoretical lens of *culturally relevant teaching* (Ladson-Billings, 1995) to try to understand the teachers' pedagogy. Through this research, we came to understand that both culturally relevant teaching and the theoretical underpinnings of the NCTM reforms had some similar orientations toward children's knowledge and thinking (Gutstein, Lipman, Hernández, & de los Reyes, 1997). Both perspectives recommended that teachers first understand children's informal thinking and experiences and then build on them. A principal distinction is that for the NCTM, the knowledge was children's mathematical thinking, while for culturally relevant pedagogy, it was cultural knowledge. We surmised that there were *potential* connections between the two distinct ways of building on children's experiences that teachers could use in the classroom.

But these connections were theoretical, and we had not seen those links being made in practice by the teachers—nor did we understand exactly what it meant to try to bring these two rather large frameworks together. Thus, in the spring of 1997, while working with Rivera teachers, I had the opportunity to take over a seventh-grade mathematics class for the last quarter of the year. I did so, not to formally conduct research, but rather to be more involved in working with students and better understand teaching and learning issues at Rivera. Also, I was interested in the more social justice-oriented aspects of culturally relevant teaching, that is, "to empower students to critique society and seek changes based on their reflective analysis" (Tate, 1995, paraphrasing Ladson-Billings). More than one teacher-research project has emerged from a teacher trying to deepen her or his knowledge of a particular situation and realizing the necessity of becoming more systematic in collecting and analyzing data (Anderson, Herr, & Nihlen, 1994). And as I taught over the next six years at Rivera, my research questions and goals became more focused and clearer.

I did not separate my teaching from my research, and I refer to my work as *teaching mathematics for social justice* (Gutstein, 2003, 2006). My goals as a teacher were two fold. In terms of mathematics, I wanted students to (a) develop *mathematical power*,[2] (b) achieve conventional success in mathematics (e.g., pass the various gate keeping tests and have the opportunities to pursue advanced mathematics and mathematically oriented careers), and (c) change their orientation toward mathematics away from a decontextualized body of rote rules to be memorized and regurgitated to a meaningful, sense-making tool for understanding and changing the world. In terms of social justice, my goals were that students should develop (a) sociopoliti-cal consciousness (an awareness of the social, political, economic, histori-cal, and cultural contexts of their lives, society, and world), (b) a sense of *social agency* (a view of themselves as people able to effect change in the world), and (c) positive cultural and social identities (youth who are strongly rooted in their home languages and cultures, and have the confi-dence and capacities to stand up for that in which they believe).

RESEARCH GOALS AND METHODOLOGY

My research goals were simple—since we have no blueprints for how to teach mathematics this way, I wanted to study the process as it unfolded, collaboratively with students and parents, and try to understand how we could collectively create conditions for students to learn mathematics so that these above goals could be reached. In this sense, it was a form of action research in that the goal of the research was to improve, in some sense, the pedagogical practice and students' learning. But my research goal was also to deepen the theoretical knowledge of the complexities of teaching for social justice in urban contexts in the United States under the conditions of increased *accountability* and high-stakes testing regimes that are embedded in, and linked to, an era of globalization and drastically increasing inequality in wealth and quality of life for the people of the world (Lipman, 2004). The ultimate goal of all of my labor is to contribute to a more just and equitable world by engaging in the theoretical and prac-tical work of creating opportunities for youth themselves to become part of the struggles for social justice and by participating in these efforts myself (Freire, 1998).

My data sources included student mathematics work and writing, jour-nals, open-ended surveys, outside observations, my own practitioner jour-nal, focus group interviews, and informal conversations with students and parents, both in and out of school. I analyzed the data using standard qual-itative research methods, coding and revisiting the data, and inductively developed themes and built theoretical propositions (Emerson, Fretz, and Shaw, 1995; Hammersley & Atkinson, 1983). As I read students' responses to a given assignment such as this one, I labeled specific blocks of text with codes. I either selected these codes beforehand (because I was familiar with certain issues or I was looking for specific things), or they suggested themselves and emerged from the data. For example, I coded specific text portions as (a) "questioning" to refer to instances in which students raised their own questions, (b) "contradictions" to refer to places where students contradicted themselves or gave double messages, and (c) "stereotype acti-vation" where it appeared the students were drawing on their own preju-dices. As well, I coded instances in the texts where students' mathematical ideas surfaced and how they interacted with non-mathematical conceptions. Some codes were specific to the topic we are studying. For example, in this assignment, codes emerged such as "drunk driving," "fighting," and "gang-banging," all of which students named as behaviors for which police arrested Latinos. Codes can also be hierarchically grouped; thus, one could subsume the above under a more encompassing code of "reasons why police arrest Latinos."

Themes emerge as one re-reads data, sees how codes interrelate, exam-ines patterns, and looks for relationships between important ideas that can help explain and give meaning to an analysis. For example, in my classes students sometimes exhibited what I refer to as a *sense of powerlessness* when they discovered an injustice through investigating data (e.g., when Marisol discovered how unequally distributed is the wealth of the world, she wrote, "Oh well, what can I do about it; I'm not rich."). However, when students expressed such sentiments, they almost always demonstrated a sense of jus-tice, anger, and outrage at the same time, and they usually connected these latter feelings to their own life experiences as members of a marginalized community that dealt with various forms of racial/ethnic and linguistic/language discrimination. Thus, the larger theme I inferred here was that these two aspects of students' identities—their sense of powerlessness and their sense of justice—were closely related. This led me to examine instances in the data where students made clear their sense of justice *with-out* feeling hopeless—and to then try to understand how to create condi-tions for them to build on that sense of justice to overcome the feeling of powerlessness and develop the capacity to see themselves as *actors in the world*, able to shape history. The coding and data analysis process led me to new insights, new questions, and new directions in both teaching and research. For a more complete description of my pedagogy, curriculum, and research methodology, see Gutstein (2003, 2006).

The class I taught in 2000–2001 was a seventh-grade mathematics class in Rivera's general track. The school has both a general and honors track (I taught in both), but students are demographically indistinguishable across tracks. Virtually all of the students are Mexican (generally first or second generation) and low income, and students live in working-class, immigrant, Latino communities. Students in the general track usually go to the neighborhood high school whose dropout rate is about 50%. The transnational community is industrious and supportive, but it also has a seri-ous gang problem and limited economic opportunities. Since many people in the community are undocumented (and subject to factory raids and "no match" letters[3]) and police are highly visible (a neighborhood corner were designated a "hotspot" of Chicago's now-defunct anti-gang loitering ordinance), the threat and experience of various forms of racial profiling was quite real to my students.

A RACIAL PROFILING PROJECT:
DRIVING WHILE BLACK/DRIVING WHILE BROWN

In February 2001, my class completed a project entitled *Driving While Black/Driving While Brown—DWB/DWB: A Mathematics Project About Racial Profiling* (see appendix for the full project). The project itself had three parts which took about a week. In Part I, students reviewed basic probability ideas. In Part II, they found, by analyzing a simulation, the percentages of African Americans, Latinos/as, Whites, and Asians/Native Americans in Chicago. In Part III, students investigated actual DWB/DWB data, and they created, ran, and analyzed their own simulation. They finished the project by writing about the issues; this was the norm for all the real-world projects in which my students investigated aspects of injustice.

To begin the project, students familiarized themselves with quantifying basic probabilities from a unit in the *Mathematics in Context* curriculum (National Center for Research in Mathematical Sciences Education & Freudenthal Institute, 1997–1998). The initial activities included ones such as finding (and quantifying) the chance of a frog randomly jumping on a black floor tile given 12 white and 4 black tiles. For Part II, I designed a simulation of Chicago's racial populations. I gave each group of three or four students a bag containing 25 cubes of different colors—nine tan cubes (to represent the whites who were 36% of the population), six red cubes (the 24% Latinos/as), nine black cubes (36% African Americans), and one yellow cube (4% Asians/Native Americans) (see Table 6.1). Without looking in the bag, each group had to pick and replace cubes, one at a time, until they recorded 100 picks. After each 10 picks, they entered the cumulative totals by color on a chart and found the percentage and fractional equivalents of each race (see appendix). At the end of their 100 picks, they had to guess the bag's contents by analyzing their data. In addition, because they recorded data cumulatively, I hoped that they would see for themselves how the percentages tended to converge over time. Since I did not tell students initially how many cubes were in the bag, that was an additional challenge, as they could only estimate. However, I found that some groups were unable to deal with that level of uncertainty and eventually told a few of the groups how many total cubes were in their bag to help them guess the numbers of each color.

When students completed the 100 picks (and some did 200), we pooled and averaged the whole class data, looked at the variance of each group's

Table 6.1.   Simulating Chicago's Racial Population

Population-Simulation	White	Latino	African American	Asian/Native American
Percent	36	24	36	4
Cubes	9 Tan	6 Red	9 Black	1 Yellow

results, and eventually opened the bags. As could be expected, the cumulative results of all eight groups more closely approximated the actual percents than any one group's data. As one student, Otilio, wrote, "The more times we picked, the closer the numbers got. The more that we pick, the better chance we have to know what is in the bag." His group had guessed that their bag contained either eight tan, eight black, three red, and one yellow, or ten tan, ten black, four red, and one yellow—both good guesses given their data.

In Part III, I introduced the DWB/DWB data. I gave students real data from a lawsuit filed by the ACLU in southern Illinois, although I averaged the data across multiple counties and then scaled them for simplicity's sake (see Table 6.2). The data I gave students were comprised of 1,000,000 motorists, 28,000 of whom were Latinos, and over a set period of time, police made 14,750 discretionary stops of which 3,100 involved Latino drivers. Students had to determine what percentage of the motorists were Latino (2.8%) and then had the task of designing a simulation that would replicate police making discretionary stops with the given population. This was challenging. Some groups used 97 tan cubes (for whites) and three red cubes (for Latinos), while two groups used 35 tan and one red, but some needed help from me in setting up and understanding the simulation. Once they did, however, they again picked and replaced 100 (or 200) times, recording the picks. The results ranged from 1% "Latinos stopped" to a high of 7% (quite high and possibly an error), and when we combined the data, we had a mean of a little over 3% for the number of Latinos stopped that resulted from the simulation. Once again, students saw the law of large numbers in action, but they saw something else as well. They saw that their simulations suggested that about 3% of the stops should be of Latino drivers, but the actual data were that 21% of the stops were of Latinos. This was, of course, the basis of the ACLU lawsuit, and the students, too, did not miss the obvious.

In the writing assignment that culminated the project, students wrote responses to the following questions: "What did you learn from this activity? How did you use mathematics to help you do this? Do you think racial

Table 6.2.   Driving While Brown Data

DWB Data	All Races	Latinos	% of Latinos
# of Motorists	1,000,000	28,000	2.8%
# of Stops	14,750	3,100	21.0%

profiling is a problem, and if so, what do you think should be done about it? What questions does this project raise in your mind?" I always tried to hold whole group discussions summarizing projects and leaving space for students to raise additional questions. It was not my intention to answer all my students' questions, nor could I. Students raised questions like, "Why do they discriminate against Latinos?" not a question one can answer easily. However, providing the space for students to pose and discuss their own questions is an important component of this form of pedagogy.

Students had much to say in response to these questions and developed their own analyses of the data. Yesenia argued that what the police were doing was justified perhaps. She wrote:

Police are maybe just stopping people because they might look suspicious, drunk, or something. They don't just stop you for anything. I think that this project was not accurate to prove anything. I mean what is it proving, it is not accurate in the cubes.

But others disagreed. Lydia's response was fairly typical:

I learned that police are probably really being racial because there should be Latino people between a range of 1-5 percent, and no, their data is 21 percent Latino people. And also I learned that mathematics is useful for many things in life. Math is not just something you do, it's something you should use in life.

Miriam wrote what seemed to her to be a reasonable response to the problem, though it may also have reflected a certain amount of naivety:

Well, I do think that Latinos shouldn't be harassed by the police, because the police's work is to have or make a good community. And too, I think that police shouldn't harass Latinos just because they are Latinos. And Latinos shouldn't let them, they should go to the police department and tell how that person was harassed just because of their race or color.

Finally, Dulcinea wrote:

> Now that you make me think about it, racial profiling really is a prob-
> lem. Why do people have to discriminate against other people just
> because of their skin color? Well, don't get me started on discrimina-
> tion, because then all of my paper is going to tell you [about that]. I
> think that there should be equal stops of white people, black people,
> Latinos, and Asian people. And the question in my head is why do
> people discriminate and why are people taught to discriminate?

DEVELOPING MATHEMATICAL POWER

One of my goals was that students develop mathematical power. In this
class (like all), students' mathematics learning was uneven. Some students
demonstrated a relatively sophisticated grasp. For example, when I asked
students to guess the contents of their bags when they simulated the racial
makeup of Chicago (the bags had nine tan, nine black, six red, and one
yellow cube), several groups guessed quite accurately and a couple were
almost exact. Other students, however, had more difficulty, including with
designing their own simulation to replicate the random stops of Latinos/as
with the data I gave them, admittedly the hardest part of the assignment by
far. And in the questions I asked students for their individual write up, I
made the mistake of including the question, "How did you use mathematics to help you do this?" in with the other questions about racial profiling,
rather than separating the questions with ample space after each one for
extended answers. Students wrote much about racial profiling, but their
comments about mathematics were fewer and not very illuminating. Of
course, that by itself is a piece of data and indicates that students felt the
need/liberty/desire to write about what mattered to them and took that
opportunity, but I received little information from the questions about
their mathematical thinking.

Although I do not have the space here to explain it, in my classes at Rivera overall, there was a significant relationship and complicated interactions between the development of students' mathematical power and their
sociopolitical consciousness and sense of social agency. Briefly, because
teaching mathematics for social justice, from my perspective, has two sets
of integrally related but nonetheless separate goals (for social justice and
for mathematics), a dialectical tension exists. Navigating the relationship
between the different goals, content knowledge, conceptual frameworks,
and perspectives is demanding of teachers and requires knowledge of both
areas. There are inevitable tradeoffs that exist that are not necessarily
problematic, but teachers need to make them explicit to students—and

this assumes that teachers themselves are conscious of them! In my own experience, I learned this sometimes in hindsight. Furthermore, there are aspects of some of the reform mathematics curricula that *potentially* support teaching mathematics for social justice. These include curricula that place students in the position of deciding on the correctness and validity of various solutions, have them create and evaluate multiple perspectives, justify their findings with solid evidence, and use real-world contexts for mathematical explorations. However, these features by themselves do not develop sociopolitical consciousness or social agency, but may play supportive roles when used in settings in which a social justice mathematics pedagogy is in place explicitly. I discuss all these aspects in depth in Gutstein (2006).

GOING BEYOND THE MATHEMATICS

The NCTM (2000) Grades 6-8 Standard on *Connections* states, "Clearly, rich problem contexts involve connections to other disciplines (e.g., science, social studies, art) as well as to the real world and to the daily life experiences of middle-grades students" (p. 274). My students, as members of working class, immigrant, bilingual families subject to race and social class discrimination, easily related this project to their lives. However, the analysis should neither stop with a surface-level understanding of something as complex as racism, nor be isolated and disconnected from the rest of the curriculum. In fact, one weakness of this particular project was that we spent insufficient time dissecting some of the complexities. For example, some students argued that police might stop drivers whose cars had broken taillights or bad mufflers, but we did not discuss why Latinos might have cars in worse condition disproportionately to the population—that is, we did not explore the correlations between race and class. Nor did we analyze the police claim that racial profiling was just good police practice because people of color allegedly commit a disproportionate amount of crime. We might have investigated this issue further and tried to understand (with and without mathematics) what are some of the social reasons people commit crimes. In fact, these further investigations are good rationales for interdisciplinary projects, like this one, that can be approached across subject areas.

Paulo Freire (Freire & Macedo, 1987) used the terms *reading the world* to refer to developing a sociopolitical consciousness of one's own experiences as well as of the broader society. Freire linked that idea to *reading the word*, or acquiring textual literacy. I also wanted my students to learn to read the *word*—in my case, the *mathematical* word—as well as learn to read the *world*, with mathematics. For me, reading the world with mathematics means to use mathematics to study and analyze social phenomenon in order to

understand issues of power in society in sociopolitical and cultural, historical context. The questions of whether or not racial profiling is racist, for
example, or what racism means and how we can fight it, are not simple
matters, and one needs a deep comprehension of history and of society to
understand and combat successfully various forms of discrimination.

Furthermore, I was not interested just that students be able to *read* the
world, I also wanted to create the conditions for them to be able to *write*
the world (Freire & Macedo, 1987). That is, I wanted them to be part of
the struggles for justice as they grew up into adulthood. Of course, one
cannot know easily what students take from a middle school mathematics
class, nor how those experiences influence their later lives. I have been
studying this question with several students from a mathematics class that I
taught in the 1997-8 and 1998-9 school years—their seventh and eighth
grades (Gutstein, 2006; Gutstein, Barbosa, Calderón, Murillo, & Nevárez,
2003). Their thoughts are that this pedagogy had an important influence
on how they see themselves and the world. As one of these students, Grisel
(Murillo) wrote in 2003, while a high school senior:

> [The class] made us all think *deeply* of otherwise superficially thought
> of issues. These deep reflections really allowed us to form opinions
> on issues having to do with injustices, stereotypes, racism, and preju-
> dice at an early age—an age when it is crucial for us to inculcate a
> sense of awareness of our world and the role we play in it.

In part, these mathematics classes encouraged students to think deeply
because the racial profiling project was not an isolated event. Throughout
my teaching at Rivera, I used real-world projects like these in which students used mathematics to investigate a variety of issues that were directly
relevant to their lives—from studying neighborhood gentrification/dis-
placement and wealth inequality in the world, to analyzing mortgage rejec-
tion rates by race and government funding priorities (see Gutstein, 2005,
for a list of issues). Students were engaged consistently in these real-world
mathematics projects even when otherwise disinterested in mathematics.
After one of the projects that we did in class, a student wrote, "You might
say that this was one of the few times that I really enjoyed math class." On a
different project, students conducted surveys on questions of their choice,
but I restricted the topics to ones that were "meaningful." After initial resistance at not being able to conduct surveys on favorite musicians (for example), we discussed what "meaningful" meant, and virtually all students took
the project quite seriously. One student, Jaime, who got himself in trouble
consistently in school, chose to work with two serious girls, and he actively
participated in the project. Together they addressed the question, "Why
Do So Many Teens Drop Out of School?" As they wrote collectively, "We

have chosen this sample because we care what other teens like us or just a little bigger think. It might be you who drops out and ends up dead or scrubbing toilets." This suggests *why* students were engaged in projects such as these. They were, in fact, meaningful to their lives, experiences, and communities, much more so than, collecting and analyzing data from paper airplanes with one or two paper clips, for example.

My data suggest that students learned mathematics and developed aspects of mathematical power, and they also changed their orientation towards mathematics—they began to see it as useful in understanding more deeply their worlds. Furthermore, through their study of mathematics in a classroom oriented towards social justice, they also began the process of developing sociopolitical consciousness. They wrote powerfully and eloquently about injustice, and their lives, futures, communities, and experiences, while using mathematics as an analytical tool to increase their comprehension of their society. As Paulina wrote at the end of eighth grade:

All my views have changed. The world before wasn't very interesting to me because I wasn't aware about all the issues that were happening. Now, math has made everyone interested in the real world because it's something new that catches everyone's attention....I think I'm able to understand the world with math. All the math problems, projects, discussions about drug testing, Chicano history, etc. have made me understand because knowing about those issues and the discussions that we did made me think of what math might be involved.

THE COMPLEXITY OF SOCIAL JUSTICE PEDAGOGY

I do not mean to suggest that teaching mathematics for social justice is easy or uncomplicated, nor that my efforts were smooth or exemplary. Many issues are involved. For example, some teachers may worry that they will influence students unduly to accept their views by showing students graphic examples of inequality. One concern is that students will figure out the teacher's supposed intentions and realize that they are "supposed" to condemn the injustice. Of course, people can be influenced, but this view belittles students, in my view, and assumes they cannot make independent judgments. As a parent of two of my students said in an interview in response to the question of whether it was possible to "brainwash" middle school students, "No, because I think they're mature enough and I have a lot of faith in them and a lot of trust that they're not gonna let themselves be brainwashed or influenced in this way" (Gutstein, in press). The challenge, as I understand it, is to provide students with the space and experiences to consider multiple perspectives and then to find and raise their *own* voices regardless of the content they study and their teacher's views—this is an important aspect of their development. While the data documenting racial disparities and other injustices are more or less "objective," the interpretations and further questions raised have to be students' own. As Rosa, one of the students to whom I taught mathematics in seventh and eighth grades wrote in 2003 when she was a high school senior, "Like I said before, I myself always questioned him [the teacher], over and over, but it was that questioning, that ability to even question the teacher that made me grow academically."

A related question some raise is whether this type of pedagogy overly politicizes education. However, like others (e.g., Bigelow & Peterson, 2002), I argue that teachers' choice of context is always political, although it may not seem so on the surface. If teachers choose contexts for their mathematics curriculum—even if they are entirely "real-world" situations like shopping, cooking, traveling, and building—without placing those situations in their sociopolitical contexts, then that teaches students several things. It suggests that politics is not embodied in everyday situations, it casts mathematics as having no role in comprehending societal inequities and power imbalances, and it provides no experience for students to be able to use analytical tools (like mathematics) to make sense out of and attempt to rectify unjust situations. These all contribute to disempowering school experiences for students, and, in my view, are thus political acts, though not necessarily conscious. Teachers may choose not to ask students to investigate real-world data that speak to social inequalities in the world, but that neither makes them go away, nor prepares students to deal with them.

An additional concern is that middle school students are too young to deal with these issues. However, that perspective misses that students who live in marginalized communities already deal with these issues and more, and we teachers who come from outside their communities need to realize that. Students have real-life experiences from which we can build curriculum so that they can learn not only the specific subject matter (like mathematics), but they can also develop deeper understandings of their life circumstances. Then there is the concern that, given the constraints of *No Child Left Behind* and testing/accountability pressures, no time is left for "extra" explorations like that of racial profiling.

Yes, those forces are real and affect everything we do as teachers. But teachers are not automatons either, and we, too, have to exercise agency. Unless we are forced to use scripted curricula, there are always spaces in which to create opportunities for students to use mathematics to develop a more profound understanding of how society functions and a sense of agency. Teachers need support and networks in *teaching to transgress* (hooks, 1994), and the work of groups like *Teachers for Social Justice* (Chicago, www.teachersforjustice.org)and the Rethinking Schools education journal (www.rethinkingschools.org) can be very useful in providing that kind of support. In mathematics, the recent publication of *Rethinking Mathematics* (Gutstein & Peterson, 2005) and the work of Marilyn Frankenstein (1983, 1995, 1997, albeit at the college level) can provide frameworks and many examples of how to involve students in learning mathematics and reading their worlds.

The final issue to which I give much thought is the issue of *teaching other people's children* (Delpit, 1988). As a white, male professional working in a Mexican neighborhood, I am quite conscious of my outsider status even after ten years of working with people in the community. I have done many of the obvious things—my Spanish is decent and improving, and I visit people in their homes and establish relationships outside of school with students and parents both. However, it is not that simple. I cannot have an insider's perspective on what it means to be the recipient of racism, even though I participate in community struggles against it. I need to be clear on what I do and do not know and realize that students are knowledgeable about their own experiences, even if they may not yet have fully articulated, systematic understandings. Two things they need are spaces in which to explore issues like racism in some depth within a supportive environment and analytical tools with which to begin to make sense out of the profound injustices that they see and experience. A concrete way that teachers can express solidarity with students, especially when they are other people's children, is to work together with them to co-create a classroom that provides these spaces and tools.

CONCLUSION

The purpose of this chapter is to present an example of teaching mathematics for social justice. This type of instruction should be rich mathematically and provide students the opportunity to construct important mathematical knowledge, like probability and statistics, and participate in developing mathematical processes as outlined in the NCTM *Standards*. From the perspective of social justice pedagogy, learning powerful mathematics *is* important in and of itself. But it is also essential in investigations such as this one precisely because without a sufficiently sophisticated grasp of the *mathematical* ideas involved, one cannot fully appreciate the *political* aspects of the situation. That is, strong conceptual understanding of the mathematics is fundamental in understanding phenomena like racial profiling, but the mathematics alone is insufficient to understand racism fully. One must also delve into the sociohistorical and political-economic context, using a variety of forms and modes of analysis, to try to reach deeper comprehension. Mathematics classrooms oriented toward justice can play a role by engaging students in substantive explorations and discussions about social structures and practices that matter in their lives, especially those that disempower and marginalize them both individually and collectively, so that they are able to work proactively to rectify the injustices they see.

My experiences attempting to synthesize these two areas—learning mathematics and reading the world—suggest that it is difficult but possible to do so, even if there is much to learn about how to do this in practice. Despite whatever shortcomings there were in the project, I argue that students need these tools with which to understand more deeply—and counter—unjust social conditions. Mathematics can be such a tool, and when students appropriate it to begin to read and write the world, the causes of equity and justice are served. As Lupe, one of my students, wrote on an open-ended survey after I turned in the final grades for her class at the end of eighth grade, and I no longer had any conventional teacher power:

> With every single thing about math that I learned came something else. Sometimes I learned more of other things instead of math. I learned to think of fairness, injustices and so forth everywhere I see numbers distorted in the world. Now my mind is opened to so many new things. I'm more independent and aware. I have learned to be strong in every way you can think of.

APPENDIX: DRIVING WHILE BLACK/
DRIVING WHILE BROWN—DWB/DWB
A MATHEMATICS PROJECT ABOUT RACIAL PROFILING

[NOTE: This is the Teacher's version]

The purpose of this project is to investigate *racial profiling*, or *Driving While Black* or *Driving While Brown (DWB/DWB)*. African Americans and Latinos/as have complained, filed suit, and organized against what they believe are racist police practices—being stopped, searched, harassed, and arrested because they "fit" a racial profile—they are African American (Black) or Latino/a (Brown). But is this true? How do we know? And can mathematics be a useful tool in helping us answer this question?

PART I. Review basic probability ideas. To understand racial profiling, students need to understand several concepts: *randomness, experiment, simu-*

lation, sample size, experimental and theoretical probability, and the law of law numbers (i.e., the more experiments you run, the closer you come to theoretical probabilities). One way to begin discussing these ideas is to have pairs of students toss a coin 100 times (the experiment) and record results, then combine the class data and have the whole class together examine how the combined data comes closer to a 50-50 split than do the individual pairs (the law of large numbers).

PART II. Find Chicago's racial breakdown. Give each group of students a small bag with colored cubes to match the racial breakdown. I used nine black (African Americans), nine tan (whites), six reds (Latinos/as), and one yellow (Asians/Native Americans) to approximate Chicago racial proportions. Do not tell students the total number of cubes nor how many of each color. Students pick one cube without looking, record its color, and replace the cube. They record the results of each 10 picks in the chart below. Each line in the chart below is the cumulative total of picks. Tell students that they are conducting an experiment (picking/replacing 100 times), collecting data (recording each pick), and analyzing data (determining from their simulation, how many there are of each color, and the total, and what are the Chicago racial/ethnic percents.

Make sure students record the fraction and percentage of each race/ethnicity for every 10 picks in the chart.

# of picks	White #	White fraction	White %	Af Am #	Af Am fraction	Af Am %	Latino	Latino fraction	Latino %	Asian #	Asian fraction	Asian %
10												
20												
30												
40												
50												
60												
70												
80												
90												
100												

Questions for each group. Emphasize *thorough* written explanations for all questions.

1. Without opening up the bag, how many cubes of each color do you think are in it? Why?
2. What happened as you picked more times, and what you think will happen if you pick 1,000 times?

PART III. Investigating DWB/DWB. Here are sample Illinois data based on police reports from 1987-1997. In an area of about 1,000,000 motorists, approximately 28,000 were Latinos/as. Over a certain period of time, state police made 14,750 *discretionary* traffic stops (e.g., if a driver changes lanes without signaling, or drives 1-5 mph over the speed limit, police *may* stop her or him but do not have to). Of these stops, 3,100 were of Latino/a drivers. Have students use what they learned in Part II and set up their own simulation of the situation using cubes (they may need more cubes, but you can let them figure this out. In my class, they either used 3 different-colored cubes of 100, or 1 of 36—this part is very difficult!). Have them pick and replace, record the data, and calculate the results of simulating 100 "discretionary" stops.
More group questions.

3. What percentage of the motorists in Part III were Latino/a?
4. What percentage of the discretionary traffic stops were Latino/a?
5. How did you set up the simulation for problem #3—how many "Latino/a" cubes and how many total? *Why* did you choose those numbers?
6. How many Latinos/as were picked out of 100 picks, and what percentage is that?
7. Do your results from your simulation experiment (#6) support the claim of racial profiling? Why or why not?

Combine individual groups' results and analyze as a whole class.

8. Individual Write up

- What did you learn from this activity?
- How did mathematics help you do this?
- Do you think racial profiling is a problem, and if so, what do you think should be done about it?
- What questions does this project raise in your mind?

End with whole-class discussion.

APPENDIX: Driving While Black/Driving While Brown—DWB/

NOTES

1. For example, police have the discretion whether or not to stop a driver who changes lanes without signaling.
2. I use the following, from the NCTM *Standards* (2000) as a working definition of mathematical power:

    ...Students confidently engage in complex mathematical tasks....draw on knowledge from a wide variety of mathematical topics, sometimes approaching the same problem from different mathematical perspectives or representing the mathematics in different ways until they find methods that enable them to make progress....are flexible and resourceful problem solvers....work productively and reflectively.... communicate their ideas and results effectively....value mathematics and engage actively in learning it. (p. 3)

3. The Immigration and Naturalization Service periodically swoops down on workplaces looking for undocumented workers, especially in sweat-shop-like factories where they are known to work. "No match" letters are ones sent by the federal government to employers stating that the social security numbers of their workers are invalid (i.e., they do not "match" existing records).

REFERENCES

American Civil Liberties Union (2003). *Urge congress to stop racial profiling.* Retrieved November 18, 2003, from http://www.aclu.org/PolicePractices/Police Practices.cfm?ID=9967&c=118.

American Civil Liberties Union (1999). *Driving while black: Racial profiling on our nation's highways.* Retrieved November 18, 2003, from http://archive.aclu.org/profiling/report/.

Anderson, G. L., Herr, K., & Nihlen, A. S. (1994). *Studying your own school: An educator's guide to qualitative practitioner research.* Thousand Oaks, CA: Corwin Press.

Bigelow, B. & Peterson, B. (Eds.) (2002). *Rethinking globalization: Teaching for justice in an unjust world.* Milwaukee, WI: Rethinking Schools, Ltd.

Delpit, L. (1988). The silenced dialogue: Power and pedagogy in educating other people's children. *Harvard Educational Review, 58,* 280-298.

Emerson, R. M., Fretz, R. I., & Shaw, L. L. (1995). *Writing ethnographic fieldnotes.* Chicago: University of Chicago Press.

Frankenstein, M. (1983). Critical mathematics education: An application of Paulo Freire's epistemology. *Journal of Education, 165,* 315-339.

Frankenstein, M. (1995). Equity in mathematics education: Class in the world outside the class. In W. G. Secada, E. Fennema, & L. B. Adajian (Eds.), *New directions for equity in mathematics education* (pp. 165-190). Cambridge: Cambridge University Press.

Frankenstein, M. (1997). In addition to the mathematics: Including equity issues in the curriculum. In J. Trentacoast & M. Kenney (Eds.), *Multicultural and gender equity in the mathematics classroom,* 1997 Yearbook of the National Council of Teachers of Mathematics (pp. 10-22). Reston, VA: National Council of Teachers of Mathematics.

Freire, P. (1998) *Pedagogy of the oppressed.* (M. B. Ramos, Translation). New York: Continuum. (Original Work published in 1970)

Freire, P., & Macedo, D. (1987). *Literacy: Reading the word and the world.* Westport, CN: Bergin & Garvey.

Gutstein, E. (2003). Teaching and learning mathematics for social justice in an urban Latino school. *Journal for Research in Mathematics Education, 34,* 37-73.

Gutstein, E. (2005). Real-world projects: Seeing math all around us. In E. Gutstein & B. Peterson (Eds.), *Rethinking mathematics: Teaching social justice by the numbers* (pp. 117-121). Milwaukee, WI: Rethinking Schools, Ltd.

Gutstein, E. (2006). *Reading and writing the world with mathematics: Toward a pedagogy for social justice.* New York: Routledge.

Gutstein, E. (in press). "The real world as we have seen it": Latino/a parents' voices on teaching mathematics for social justice. *Mathematical Thinking and Learning.*

Gutstein, E, Barbosa, M., Calderón, A., Murillo, G., & Nevárez, L. (2003). *A Freirean approach to learning mathematics in an urban Latino/a middle school: Examining the long-term influence.* Paper presented at the Annual Meeting of the American Educational Research Association, Chicago, IL.

Gutstein, E, & Peterson, B. (Eds.) (2005). *Rethinking mathematics: Teaching social justice by the numbers.* Milwaukee, WI: Rethinking Schools, Ltd.

Hammersley, M., & Atkinson, P. (1983). *Ethnography: Principles in practice.* London: Tavistock Publications.

hooks, b. (1994). *Teaching to transgress: Education as the practice of freedom.* New York: Routledge.

Ladson-Billings, G. (1995). Toward a theory of culturally relevant pedagogy. *American Educational Research Journal, 32,* 465-491.

Lipman, P. (2004). *High stakes education: Inequality, globalization, and urban school reform.* New York: Routledge Falmer.

National Center for Research in Mathematical Sciences Education & Freudenthal Institute. (1997-1998). *Mathematics in context: A connected curriculum for grades 5-8.* Chicago: Encyclopedia Britannica Educational Corporation.

National Council of Teachers of Mathematics. (2000). *Principles and standards for school mathematics.* Reston, VA: Author.

Tate, W. F. (1995).Returning to the root: A culturally relevant approach to mathematics pedagogy. *Theory into Practice, 34,* 166-173.

CHAPTER 7

SUPPORTING MATHEMATICAL THINKING

A Collaborative Inquiry

K. Ann Renninger
Swarthmore College, Swarthmore, PA

Susan Stein
Portland State University, Portland, OR

Judith Koenig
Retired from Boulder Valley School District, Boulder, CO

Art Mabbott
Seattle Public Schools, Seattle, WA

How do I know that I am right? We ask this question as we make decisions in the classroom. To support our students' mathematical thinking, we also encourage our students to ask, "How do I know that I am right? In this chapter, we describe findings from a collaborative inquiry that focuses on these questions. Our thinking about these questions began with the

Teachers Engaged in Research
Inquiry Into Mathematics Classrooms, Grades 6–8, pages 119–146

shared experience of participating in the Math Forum's Bridging Research and Practice Project (BRAP).[1]

The Math Forum's BRAP project was a three-year National Science Foundation sponsored project in which teachers from all over the country and Math Forum (mathforum.org) staff members thought together and wrote a videopaper about math problems, problems that others have tackled and reported in their research, and dilemmas that we face in the classroom. Following the Math Forum BRAP project, we have continued to talk online and offline about our work in the classroom. In this chapter, four of us share insights from these discussions with illustrations from our written reflections. These reflections include many references back to our work together as part of the Math Forum BRAP project, and our work with our students during and following the project.

We bring a wide range of school experiences to this discussion. We have experience working with students and colleagues in both public and private schools that are located in either the inner city or the suburbs in vastly different regions of the United States. We have worked with students who differ in ethnicity, race, gender, and socio-economic status, as well as in age.[2]

Comparing our reflections and realizations following the completion of the Math Forum BRAP project, we find that despite our different students and contexts, our experiences have marked similarities. We are each working to support our students to think and do mathematics. We are also thinking in new ways about our students' strengths, needs, experiences, and interests. We know that our students should be able to do mathematics, although some need more support than others to ask and then pursue their own "curiosity" questions about mathematics.

Using the methods of action research, we extend discussions about how to support mathematical thinking that were included in the videopaper that we wrote with other Math Forum BRAP participants (mathforum.org/brap/wrap). At times, the research on which we report takes the form of the research articles that we have read. This research also involves our own informal case studies and retrospective recall, as well as systematic video analysis of the way in which we (and others who participated in the Math Forum BRAP project) taught the same non-routine challenge problem to our classes and engaged in onsite and virtual discussions about these videos, our work in our classrooms, ourselves as learners, and the research that we read. We start by providing some background about the beginnings of our continued collaboration and then describe our present inquiry.

BACKGROUND

As participants in the Math Forum's BRAP project, we worked together in face-to-face workshops and online to research and write a video paper, *Encouraging Mathematical Thinking: Discourse Around a Rich Problem*, that continues as an online discussion of mathematical thinking (see mathforum.org/brap/wrap). At the first BRAP workshop, we each brought an activity to share. One of these activities was the cylinder problem:

> Take a rectangular sheet of paper and roll it to form a cylinder. The paper has become the lateral surface area of the cylinder. What if you rolled the paper from the other side to form the cylinder? It would have the same lateral surface area, but would it have the same volume? (see http://mathforum.org/midpow/solutions/solution.ehtml?puzzle=54 for a middle school version of this problem)

Our discussions of our own solution strategies for the cylinder problem led us to some of our first realizations about the role of language, tools, and collaboration in mathematical problem solving. As Judy noted:

> Even though I had worked the cylinder problem many times and thought I knew its possibilities, when we began our work together I found myself excited that this one problem, simple yet elegant, became a basis for talking about mathematical thinking and discourse. (JK, notes)

By working together on problems like the cylinder problem and interacting as learners of mathematics, we were able to think about the types of supports that we needed, also about those that our students might need.

The opportunity to solve problems and talk with our colleagues about their work with their students on the same problems led to powerful learning. The experience of solving problems together also led us to consider the question: How do we know that we are right? As a group, we hypothesized that it may be only when students are engaged in solving problems and asking their own questions about those problems that mathematics learning is occurring. In order to think further about promoting student learning and to compare our methods, we each decided to videotape ourselves teaching the cylinder problem in our own classes when we returned home. In order to inform ourselves about others' work on these questions, we all also contributed suggestions about articles to read as a group. Our conversations naturally turned to the links among these articles, our work with rich mathematics problems, our insights from working with problems together, and our classrooms.

We began by reading: (a) Schoenfeld's (1987) chapter about the way that metacognition could support students to think about the math, (b) Resnick's (1988) chapter about the importance of recognizing that students may do best to view mathematics as an ill-structured discipline, (c) Lampert's (1986) article describing her work to help students generate and testing hypotheses when connections between concrete materials and computational practices are made salient, (d) Atkins's (1999) article about the importance of really listening to students, and (e) Goldsmith and Shifter's (1997) chapter suggesting that mathematics reforms necessitated changes in the way in which mathematics is taught. Interestingly, one of the last books we read together as the BRAP group was Stigler and Hiebert's (1999) volume, *The Teaching Gap*. In many ways, our experience as participants in the BRAP group was the same as participating in a lesson study.

In the BRAP videopaper (The Math Forum BRAP Group, 2000), we described our findings about how and when to consider discourse a basis for encouraging students' mathematical thinking in classrooms and for supporting professional growth. We focused on methods that can be used to help students become engaged in inquiry and take more responsibility and initiative in their learning. We also used video clips from our own classrooms to illustrate the interventions discussed. These *artifacts of practice* (Ball & Cohen, 1999) allowed us to recognize that:

> … when we carefully examine the work of our students and provide them with opportunities to think and discuss among themselves, our students and we are actively learning. Focusing on the same problem in each of our classes enabled us to think about the variance that our own interaction with our students contributes to their abilities to think mathematically. Although we teach students at different levels of mathematical knowledge, we find that the issues we face as teachers are similar: how to help our students question, how to teach problem solving, how to provide students with the tools they need. We have also learned that the solutions to these problems are many, and that they become clearer when we have the opportunity to think them through with our colleagues…. This process not only enhanced our abilities to converse, it also meant that we documented alternate solutions to similar problems. It has pushed us to try others' suggestions and to revise plans that we'd thought we knew worked, in turn

encouraging us to push our thinking about mathematics teaching.
(The Math Forum BRAP Group, 2000)

Findings reported in the videopaper suggest that teachers *and* students
need to have opportunities to (a) assess what they do and do not understand in a situation, (b) have experience with alternative solution strategies, and (c) identify tools that allow them to teach and learn. Students
and teachers need to develop a repertoire of different ways to approach
mathematical or pedagogical problems. Further, they need to be able to
decide what strategy might work best and, by reflecting on the result, evaluate whether it helped them make sense of the problem. As a result, we
encouraged our students to explain and justify their methods and solutions, by asking themselves, "How do I know that I am right?" At the same
time, in our practice as action researchers and classroom teachers we also
wanted to know, "How do I know (my teaching decisions) are right?"
These questions continue to inform our work together.

PRESENT INQUIRY

In this section, we share our more recent thinking about supporting our
students to think mathematically. By mathematical thinking we mean strategies such as problem-solving approaches (e.g., model building, trial and
error, elimination), mathematical processes (e.g., making conjectures,
explaining, justifying, revising), and reflection necessary to understand
quantitative and spatial relationships and symbolic forms (The Math
Forum BRAP Project, 2000; Renninger, Farra, Feldman-Riordan, 2000;
Rothschild, 1997; Schoenfeld, 1992).

The case of one of Susan's students is used to describe the complexity of
supporting mathematical thinking. This case is intended to be a kind of
"every student" example. We all have had students whose strengths and
needs led us to take risks and caused us to question our pedagogical decisions. In this case we describe interrelated mathematics inquiries—Susan's
inquiry into teaching and her student's inquiry into learning. Following
discussion of the student's case, we point to the ways in which reflecting on
our own learning in the Math Forum BRAP project helped us to think
about teaching and learning mathematics. Following this, we point to particular aspects of the project that supported us to change our thinking
about mathematics and pedagogy.

Specific examples have been selected from our notes for purposes of
illustration. While our recollections may be in the voice of one or the other

of the authors, we each have had parallel experiences. We used methods of discourse analysis (Gee, 1999) to identify examples to share that were representative of our experiences generally. Importantly, we all have had experience with a variant on any given example cited. Data with which we worked included our notes and reflections on the BRAP project as well as our thinking, correspondence, and writing following the project. Specifically, we share written reflections on videotapes of ourselves teaching the cylinder lessons, email exchanges (fundamental to the way stayed in touch between meetings), notes from our meetings, and our reflections on BRAP meetings and their impact on our classroom practice.

We begin by focusing on our own experience of the BRAP project and its impact on our practice. These data were shared and examined for: (a) ways of thinking about and valuing knowledge, or what Gee calls semiotic building, (b) possibilities, or what Gee calls world building, (c) the context of activity, or what Gee calls activity building, (d) change in a person and group's sense of identity, or what Gee calls socio-culturally situated identity, (e) change in relationships with content such as mathematics, pedagogy, and so forth, or what Gee calls political building, and (f) comparisons between interactions and exchange, or what Gee calls connection building. These analyses provided a tool for thinking further about the opportunities our students might need in order to think mathematically and how we might address those needs through our practice.

Inquiring Into Teaching and Learning Mathematics: Shayla, an Every Student Case

We present the case of Shayla in order to highlight the decision making that teaching involves—the kind of decision making that leads us to wonder, "How do I know that I am right?" Each of us has struggled to know whether we were right in working with a "Shayla." Susan's notes are used to identify the issues that we have come to consider basic to the discussions of how to support students to think mathematically. Reflecting on her work with Shayla (a pseudonym), Susan wrote:

> I remember trying to edge Shayla to try using the algorithm the class had just derived following small group work. The small groups had solved and then presented various methods and solutions for problems similar to the following [see Table 7.1]:
> Based on previous work in the class, the students were encouraged to use percents to create their comparison.
> Shayla was an eager and willing student, but was having a difficult time making sense of percents and proportion. After a bit of back

Table 7.1.   Participation in Exercise Walking.

Activities	Males	Females
Exercise Walking	21,054,000	43,373,000
Total in Group	111,851,000	118,555,000

Problem: Use the table above that shows national data on exercise walking. Write a statement comparing the number of males who walk for exercise to the number of females who walk for exercise (Lappan, Fey, Friel, Fitzgerald & Phillips, 1998, p. 21).

and forth, I asked her why she preferred the method that her group had come up with to the more efficient algorithm other groups had presented. To me, her approach using many steps seemed so convoluted that it was barely useful. Her group used the numbers just as they appear in the table (no estimation, no using smaller numbers). They used was a lot of, to my eyes, unnecessary arithmetic manipulation. However intricate, I knew that their method as well as their solution was correct. I also knew that Shayla's past relationship with mathematics was tenuous, at best, but she was working steadily and I wanted to respect this. She replied, "Because it makes sense to me. I'll keep trying to figure out the shortcut, but right now I know I can do it my way." I decided to follow her progress. Over several weeks, Shayla did pare down her method and began to use the algorithm effectively. If I had introduced the algorithm instead of having the students derive it first in groups and then as a whole group, I doubt that Shayla would have been able to make sense of it. (SS, notes)

For both the teacher and the student in this type of situation, a key question is the one on which the BRAP project focused: "How do I know that I'm right?" Shayla is working to understand percent and proportion. Susan finds herself weighing how to support Shayla's developing knowledge and how to help her build on it. Shayla is making sense of a solution strategy, while Susan is deciding on a pedagogical one. If the goal of our work as teachers is student learning, our goal is being met. Shayla is assuming responsibility for her learning and is exerting effort to figure out the concept being studied. That she is a weak student makes her effort to develop her own method all the more remarkable.

Susan's pedagogical choice is to trust Shayla to know what she needs at that moment, and to keep watch for ways to help her deepen her understanding and steer her towards a more direct mathematical solution method. We may need to let go of the fact that it takes Shayla several weeks to process and understand a concept and adopt a more traditional algorithm. What was to have been covered in a few days may take Shayla additional time. If we focus on Shayla's determination, we can assume that if Shayla runs into difficulty, she will ask questions. Furthermore, if Shayla asks questions that emerge as she works, Shayla will be more likely to better understand percent and proportion than if she had simply accepted the traditional algorithm as her method. In fact, it is also likely that if she had simply adopted the algorithm, Shayla would not have been ready to move on to the rest of the concepts to be covered during the marking period since she would not have had as solid a conceptual understanding of percent and proportion (Ball, 1993; Lampert, 1986).

That Shayla intends to make sense of the "shortcut" is significant. Deciding that she is headed in a direction that will be fruitful is a kind of educated risk that Susan, the teacher, takes. There are at least three types of information that inform Susan's decision making in this situation. First, Susan recognizes that Shayla has a plan. Even if Shayla runs into difficulty, this plan is likely to lead to learning since Shayla will build on what she does understand to ask questions and continue to work. Second, even though Susan finds Shayla's approach convoluted, the sense making that it enables is critical. As Resnick (1988) points out, students need to understand that mathematics is *ill-structured* before they are able to appreciate its regularities and elegance. Rather than viewing mathematics entirely as established procedures and accepted proof, students and teachers need to recognize that mathematics also includes active investigation and problem solving (Cooney & Shealy, 1997). They are then able to recognize that exploration and using even unsophisticated, intermediate methods can support and develop understanding. Sense making is essential to thinking mathematically (Ball, 1993; Lampert, 1990; Schoenfeld, 1991). Convoluted or not, the teacher needs to make sense out of the student's work. Is the student's work equivalent to the accepted/preferred method? Or did she make it to the correct solution by a series of compounding errors? Third, Susan's knowledge of the curriculum informed her decisions. Susan knew Shayla was working with a coherent curriculum and that as a class they would be continuing to work with percent and proportion a while longer. Shayla would have the time to revisit the underlying idea(s) and thereby rethink and refine her algorithm.

Reflections on Supporting Shayla to Think Mathematically

Shayla's case illustrates a situation in which the teacher cannot know the outcome of her decision ahead of time; she must make an informed choice about how to proceed. In this case, Susan trusted Shayla to continue to make sense of the concepts of percent and proportion and refine her own way of thinking about them. Susan's decision was informed by prior work with Shayla and her sense of the alternatives. She could tell Shayla to use the standard procedure, whether or not she understood the concept or the algorithm. With practice, Shayla would probably have been able to reproduce it on a test. However, this may mean that Shayla would not look for her own methods in the future. Another risk was that Shayla might be able to reproduce the algorithm with familiar problems, but be unable to use it in new contexts. Susan wanted Shayla to *own* the procedure because she had constructed its meaning for herself. Only then would the algorithm become part of Shayla's repertoire for flexibly solving problems having to do with percent and proportion. Susan made a choice based on what she valued for her students. She decided to wait and watch Shayla as she worked with future percent or proportion problems.

As Susan's work with Shayla suggests, supporting the development of mathematical thinking is multifaceted. It includes student characteristics (e.g., interest, experience, strengths, needs), teacher characteristics (e.g., willingness to wonder, ability to say, "I don't know," resourcefulness, understanding of student thinking and its development, knowledge of content to be learned), and criteria or means by which the material to be learned will be assessed (Brown, Bransford, Ferrara, & Campione, 1993). Providing that we recognize that a student's method is correct (albeit convoluted), we can adapt the lesson based on our knowledge of our students and the mathematics. We can choose to question or probe a student's method, providing the student with feedback that meets his or her strengths and needs (see discussion in The Math Forum BRAP Group, 2000). We can also establish classroom expectations that encourage students to reflect on and communicate their understanding. Thus, while a student is gradually adding to his or her repertoire of mathematical solution strategies and refining an approach to problem solving, we can draw on our own repertoire, choosing a teaching strategy to foster that student's mathematical thinking.

Content knowledge allows us to redirect and/or reword in order to support mathematical thinking. However, if we do not know whether a method is correct, we may lead a student to make an unnecessary or inappropriate correction. Alternatively, if the answer is correct, s/he may fail to note illogical methods. In addition to being constrained by limits of our own content knowledge, we can be further constrained by the circumstances of the situation (e.g., limitations imposed by the school context, such as time constraints in the schedule and constraints imposed by standardized testing). In addition, we may be constrained by limited familiarity with models of how to work with students who are struggling with concepts, limited knowledge about a student's mathematical experience, limited time to process new content knowledge or curriculum, and limited time to process potential instructional changes. When teachers make a shift from one curriculum to another, for example, it can sometimes take 3–5 years for them to really understand the change in which they are involved and to make the requisite methodological changes (Reinhart, 2000).

Susan's (and our) goal is to help students learn to question, reflecting on what they know as a way to develop their mathematical thinking. Shayla indicated that she was working on understanding and this suggested that she would also be ready to ask questions as she had them. Shayla may have been supported to take initiative by a classroom that included group work and encouraged students' to ask their own questions and make connections to mathematics through modeling and apprenticeship (Collins, Brown, & Newman, 1989; Cognition and Technology Group at Vanderbilt University (CTGV), 1997; Webb & Palincsar, 1996). In this environment, Shayla may have felt comfortable to work at the pace she was able (Noddings, 1992). Although we can only identify the possible configuration of conditions that supported Shayla in this instance, we can be pretty clear that she was supported to ask and pursue her own "curiosity questions" about mathematics independently (Renninger, 2000; Renninger, Sansone, & Smith, 2004; Sansone & Berg, 1993; Zimmerman & Risemberg, 1997). In contrast to procedural questions, curiosity questions lead a student to make the kind of connection to mathematics that contribute to re-engagement, interest development, and real learning (Renninger, 2000).

REFLECTIONS ON OUR OWN LEARNING

For us, the process of thinking together about how students learn is infused with our shared experiences as learners during and following the BRAP project. Having explored how Susan (and we) work to support students like Shayla to think mathematically, we turn to other ways that collaborative inquiry supports our pedagogical thinking, guided by our question, "How do I know that I am right?" Reflecting on our own thinking and learning during and following the BRAP project allowed us to claim new strategies as our own and added new mathematical processes and pedagogical choices to our repertoires.

In this section, we describe findings from review of our own and each others' reflections about our own learning. We identify three types of changes in our relationships with math and pedagogical content and compare our reflections and exchanges over time. First, we point to conditions that are likely to be in place if strategies and reflection occur. These conditions may include other people with whom to think (Cobb, 1995), models of how to approach problems with which we work (Schoenfeld, 1987), alternate possibilities for problem solution (Ball, 1993; Lampert, 1986), the requirement that an effort be made or attention be directed to a task (Marx, Blumenfeld, Krajcik, & Soloway, 1998), recognition that what needs to be accomplished is worthwhile (Eccles et. al., 1983; Wigfield & Eccles, 2002), a succession of experiences that build on each other (Bruner, 1966), positive feelings (Mandler, 1989), a sense of possibility (Markus & Nurius, 1987), and tasks that allow multiple points of access (CTGV, 1997).

Second, we suggest that the kind of real learning we seek for ourselves and our students is best supported by not over-specifying what needs to be learned, who needs to be learning in which way, and/or when this learning needs to occur. Finally, we observe that we may not really be able to take stock of all of the ways in which we are learning and its impact on what we are doing until we have alternate examples with which to work. This argues for using group and project-based work in our classrooms and calling attention to different strategies. It may mean that we really need to learn from thinking with and watching our students as well as one another. We also need to bring this learning to our practice.

In order to describe ourselves as learners, we again draw on notes from our written reflections for illustration. First, Judy describes making the transition from teaching first graders to seventh graders, and her experience of not knowing how to work with the language of reform practice and its suggestions. She also points to the role of collaboration in helping her to connect the NCTM *Standards* to classroom practice. Collaboration also provides a basis for thinking about how to teach many groups the same problem. Then, Susan describes the sharing videos as part of the BRAP project and its relevance to her question about how to encourage student participation. Her reflection underscores our own realization that experiences can and do have different meaning, and impact, on learners. Finally, Art's notes describe the evolution of his thinking about sharing as a way to support students to use technological tools to develop their understanding and an awareness of their repertoire.

Following this, we point to aspects of the Math Forum's BRAP project that supported our own mathematical and pedagogical thinking and learning. Interestingly, we each also deepened our understanding of technology through the project as well. We used various forms of technology to communicate, and the videopaper that we developed was a first in the field. Because technology was a tool, and not the focus of the workshop, our reflections tend to center our awareness of changed thinking and practice. Technology made this a possibility.

Judy: How Do I Teach the *Standards*?

Judy was searching for help the first time that she read the *Professional Standards for Teaching Mathematics* (NCTM, 1991). At about the same time, she was moving from teaching first grade to teaching seventh grade mathematics. Judy was concerned that while her first graders had been open to everything, middle school students might be guarded. She had found that first graders could explain their ideas more fluently than they could write about them. Working with first graders meant more hands raised than she could typically handle in a single day. She was aware that middle school students, like first graders, are capable of learning a great deal, but they can be reluctant to speak out in class for fear of making an error in front of their classmates. Judy felt confident that she could present lessons that would provide the groundwork for conceptual understanding, but she wondered how she could get her students to explain their thinking to their peers. Since she believed that students explore mathematical ideas by working on problems and sharing solution methods, or problem-solving strategies, she wanted her classroom to be an environment that included this kind of peer work. What she read in the *Standards* sounded good to her:

The teacher of mathematics should orchestrate discourse by:

- posing questions and tasks that elicit, engage, and challenge each student's thinking;
- listening carefully to students' ideas;
- asking students to clarify and justify their ideas orally and in writing;
- deciding what to pursue in depth from among the ideas that students bring up during discussion;
- deciding when and how to attach mathematical notation and language to students' ideas;
- deciding when and how to provide information, when to clarify an issue, when to model, when to lead, and when to let a student struggle with a difficulty;
- monitoring students' participation in discussions and deciding when and how to encourage each student to participate. (NCTM, 1991, p. 35)

Judy found it difficult to figure out how to put the suggestions from the *Standards* into practice in her classroom, however. She wanted to be able to create middle school tools to add to her teaching repertoire. She wondered:

How should I proceed and how would I know that the strategies I chose were successful? Although approximately 30 students per class surrounded me, I was virtually teaching in isolation. I would need to do it all by myself. I could develop a strategy, implement it, observe how it unfolded in the classroom and then reflect on its impact on improving student discourse. This was difficult because I was part of the process. How could I know that it was working, how would I know that I was "right"? (JK, notes)

Like Shayla, Judy needed support to work out her own approach for implementing others' suggestions. She did not want to be told what to do, but rather she sought out opportunities to make sense of the *Standards* and their application. Judy found a succession of workshop experiences really helpful.

As was also the case for Shayla, the contexts that provided Judy with support were ones that (a) included other people thinking with her, and (b) provided models to approach the problems on which she was working. Importantly, Judy needed more than a single workshop experience to make sense of the *Standards*. To persist in her efforts, she may have needed the requirement that she shift from teaching first to seventh grade, as well as some recognition that the *Standards* were useful and considered important to her department chair and administrators.

Judy: How Do I Teach Many Groups the Same Problem?

In her notes, Judy reported that she found the online discussions between the face-to-face BRAP workshops meaningful because she had met the people with whom she was talking and could picture and hear them, as she "talked" to them. She "felt safe to express my innermost doubts and questions... reflections that she would not share with students or their parents or with my administrator." In an early posting, she questioned whether she was doing everything she could to support her students to think mathematically:

A few of my students commented at the close of class, "this class really goes by fast." Now, I could take that as a compliment meaning that the class is so interesting, it really doesn't drag—and I do comment to my students, "That's because you LOVE math so much and it is so much fun, it seems to fly by." However, in my own reflections, I'm concerned that the class does fly by and I wonder:

- Did I touch base with enough students?
- Have I checked for understanding?
- Have I answered enough questions?
- Have I given them time to be recognized as an integral part of the class?

- Has there been enough time for students to "talk and think" mathematically?
- Have I covered the content in enough depth?
- Do I have the time needed for curriculum as well as students?
- Have I taken enough time to give positive feedback and encouragement, especially with my more needy students? (JK, online discussion)

Judy was pretty clear that the more students could learn from each other, the more they would expand their conceptual understanding and valuing of mathematics. Additionally, the more they learned to value group problem solving and reporting, the better able students would be to incorporate collaboration into their repertoire for dealing with real life problems. As she planned how she would teach the cylinder problem, Judy recalled that in the past she would have had students work on solving a problem and then have them do group presentations. After the first few groups presented, however, she always found that there was not much more to add. She wondered how she could change this so that every student would have something to contribute.

For the video lesson, Judy decided to devise a variety of questions that could be given to each group as they were completing their visual presentation on poster paper. In this way, each group would have something unique to contribute after they had completely solved the problem. By changing the variables on the axes for different groups, Judy was able to expand the group presentations. For example, some groups graphed volume vs. radius of base whereas other groups graphed volume vs. area of the base. Each group was asked to present their work to the class and (a) describe how they decided to set up their graph and table, (b) explain how they decided which variable to use as the independent variable, (c) tell how they set up their scale, and (d) provide information about the general form of the equation. Since groups each reported on a different interrelated problem, there was more discussion. Judy also felt that using different perspectives to look at the same concept would stretch their mathematical thinking and understanding.

Susan: How Do I Encourage More Student Participation?

As Susan watched the video of Judy teaching the cylinder problem, she had reflected and focused on the way in which Judy facilitated the discussion of the students' work. In her notes about viewing the video, she writes:

I was struck by the way in which Judy elicited responses to her questions. When she asked her students a question, many students raised their hands. Judy called on each one in turn to share his/her response. Judy's only replies were to solicit additional ideas. She neither praised nor questioned what each student said. Her body language and facial expression were neutral, although warm and friendly. Only after all of the students who wished to contribute had finished, did she begin a discussion. I knew that it was rare for more than a few students at a time to raise their hands in my classes. After watching the video, I asked her about this.

During our discussion, I was impressed that Judy could tell me what she was thinking about as she listened to each student's response. She described the deliberate choices she made about what to say. I almost always thought on my feet and just did what felt right. Often, in retrospect, I thought I'd made the right choices, but not always. Certainly, I did not carefully think through all the alternatives each time I made a choice in the classroom. I was impressed by Judy's capacity to reflect on her teaching while she was in the middle of doing it.

My practice was affected by watching a different teaching strategy and also by Judy's ability to talk about what she did and why she did it. I consciously tried out her practice of eliciting all possible student responses without comment. I discovered that when a range of possible answers were on the table, students were more likely to question one another, compare alternatives, and attempt to understand different approaches. They also were more likely to try to answer the question: "How do I know that I am right?"

Over the next few months, as Judy's practice became my habit, I saw increased student responses in my own classes. I found it wasn't that hard. Over time, my students' discussions shifted from analyzing, clarifying, or defending one idea to comparing and contrasting several. Not surprisingly, our classroom conversations became much richer too. (SS, notes)

Art: How Do I Broaden My Repertoire and the Repertoire of My Students?

Working with the cylinder problem led us to the challenge of deconstructing the question, "How do I know that I'm right?" Only after some time passed did we realize that we were using a number of different tools to work with the problem. Some BRAP participants grabbed graphing calculators, plotted points and then looked at the graph in order to draw some conclusions. Others went to paper and pencil and used their algebra skills to create an equation. Some used spreadsheet software to look for patterns and graphs of the results. Yet others used the dynamic geometry software, *Geometer's Sketchpad*, to draw pictures. Later, we learned that the two people who had elected to work with paper and pencil also spent some time watching everyone else work. In the discussion that followed, they said they had no clue what the rest of us were doing. Their lack of experience with technology prevented them from thinking of it as just another tool to be used to aid their understanding of the problem and they were, as they reported, simply awed.

In his notes, Art Mabbott reports what happened following the first BRAP workshop. As his middle school class was exploring circles, he asked his students about the measure of angles formed by a tangent and a chord drawn to the point of tangency: "What is the measure and how do you know you are right?" He said he had not yet consciously integrated the question about knowing whether you were right into his teaching, but he sees now that he must have been thinking about it.

When I asked the class for any guesses, one of my students came up with a conjecture that it would be half of the measure of the intercepted arc. When I pushed him to explain why he believed what he believed, he started to wave his hands at his diagram on the board but then stopped and announced that with *Sketchpad*, he could simply show us why this problem was just the extension of (limit of) the inscribed angle problem—if you turn one of the chords into a ray and slide it around until it becomes tangent to the circle, you get the tangent/chord situation.

While many of my students needed another explanation, this student had integrated *Sketchpad* into his thinking/problem solving tool kit and could use it when he needed to help him solve a problem. He knew (a) that he was right and (b) that he could convince the rest of his classmates that he was right with the use of a tool—the technological tool afforded possibilities for demonstrating and thinking that regular chalk and board, or paper and pencil, did not. (AM, notes)

Art talked to his students about the importance of developing a toolkit for thinking and problem solving—for figuring out if you know you are right—although he did not at that point realize that his decision to do this could be linked back to discussions with the BRAP group. Certainly, calling experiences to awareness is likely to further thinking about what we are learning (Dewey, 1933; Rodgers, 2002; Schoenfeld, 1987). Through later reflection about his use of the question generated at our BRAP workshops, Art realized that, just as he asked his students to expand their toolkits, making the question part of his teaching practice had broadened his own repertoire.

The toolkit is a term Art, and his peers and students used at Newport High School. He does not think that they coined it, however. For the teachers and students at Newport, the notion of the toolkit started with a trigonometry class and the equation: $Y = A\sin (B(X-C)) + D$. They decided to have the students first look at the $Y=\sin (X)$ as the generating function and then A, B, C, and D were the tools to adjust the function by compressing or expanding and shifting left or right, or up and down. Then, they carried this thinking back to the general function $Y=f(X)$, and could look at the same adjusters $Y=Af(B(X-C))+D$ in the same way. This helped the students put the general quadratic into a form that then allowed them to pull out the vertex and talk about shifts and compressions/expansions. The students were creating a set of tools that would fit into their tool kit to help them analyze any function. After this experience, he began working with his students to start to create toolkits earlier—when exploring signed numbers. Rather than learning one set of rules for adding and a different one for subtraction, they learned tools for adding and a single tool to change subtraction into addition. Similarly, in Geometry, he worked with his students to add new conjectures/theorems to their toolkit as they were proven. The process of developing a toolkit, by definition also led them back to the question: How do I know that I am right? Toolkit, used in this sense, is like our notion of a repertoire of problem solving strategies.

ACTIVITY THAT SUPPORTED US TO LEARN

Three types of activity supported our learning and work together. These included technology, development of a shared language, and collaboration around rich mathematics and pedagogy as part of the Math Forum BRAP project.

Technology

When used with reflection, technology is a tool for stretching understanding and supporting learning. Our work with the BRAP project provides many examples: the use of (a) *Geometer's Sketchpad* or calculators to enhance our work with a problem, (b) video to capture our own classes, (c) videos from the Third International Mathematics and Science Study (TIMSS) of teachers from Japan, Germany, and the United States, (d) online discussions as a resource for continuing to collaborate at a distance, and (e) the videopaper as a means for expanding the reach of our discussion. The videopaper is a tool to disseminate our insights and invites others who were not part of the BRAP project to think with us about mathematics teaching and learning. As Art pointed out in his notes,

> Before we each returned to our classes to present the cylinder problem, we had extended the problem ourselves in several directions in the process of working on it. Given the 8 1/2 inch by 11 inch piece of paper, how many different cylinders can you make? That is, keeping the area fixed, how many different ways could we slice up the paper to easily form rectangles that can then be formed into cylinders? And, once you have this family of cylinders, is there a maximum cylinder—which one has the largest volume? By cutting and re-taping we were able to collect an entire family of cylinders and then collect data and create hypotheses. But how could we test it out? How could we know that we were right?
>
> The lesson had us collecting experimental data from our family of cylinders. Some of the team had their classes then analyze those experimental data and use a regression tool on their calculators to look at these data. I chose to have my students calculate the volumes of the cylinder family members by hand (or with a calculator) and compare the results. Since the length of the rectangle would be turned into the circumference of the cylinder base and this radius would be the radius of the circle needed to determine the area of the base of the cylinder which in turn was necessary to complete the calculation of the volume, we needed to find the length of the rectangle, the radius of the circle, the area of that same circle and then the volume of the cylinder—four separate calculations. And, an error on any one of these would throw off the rest, so there was a real need to be accurate.
>
> But as we progressed through the activity, I began to notice that while the calculations were not difficult, the number crunching became tedious and my students were prone to calculation errors. Since the goal of the activity was to explore a deeper problem—is there a cylinder with a greatest volume?—rather than practicing finding a radius from the circumference and then area, and finally volume, I chose to use a little technology.
>
> We built a spreadsheet that would take our given rectangle dimensions and calculate our needed cylindrical values—the radius of the

circle, its area, and finally the volume. We first calculated the values
that the groups had been working on—the "nice" integer values—to
validate student answers or to help them understand where they were
confused and the source of any miscalculations. Later, we were able
to quickly and easily add values (add cylindrical family members) to
our data pool and fill in the graph a little better to help with the basic
understanding and this, in turn, allowed the conversation to take on
depth. (AM, notes)

We used technology to extend the way our students and we engaged
with problems. Moreover, we found that technology was most useful when
it provides insights that one is not able to get in some other way.

Developing Shared Language

Reading and thinking together about students' learning mathematics provided us with a language that we could use to talk about tools such as technology. At the start of the project, only Ann had a background in the
research literature. Initially, she helped to choose the articles and chapters
BRAP participants read. Over time, however, we all began to seek out and
share articles. Even though we all came from different classrooms—some
had very large classes and some very small; some were public school classrooms, some were private, and some were online; some were middle school
classrooms and others were in high school or university settings—our common ground and shared language emerged from the work we did together
to solve problems and make sense of the readings.

Susan recalls that when she first joined the project, she could barely find
time to keep up with the NCTM journals she received. However, she found
that discussions of readings led her to some of her most potent learning
during the project.

As a practitioner unfamiliar with education research, initially I was
pleased just to discover that I could read and understand the articles.
They made sense, made points that I recognized from practice, and
helped me to articulate my own experiences. In addition, I found
some of the ideas challenged me to improve my teaching, while showing me avenues for doing so.

Seventh graders usually came to my classroom at the start of the
school year expecting a teacher who would tell them when their
answer was correct. They expected to be told when they were on the
right track, and to be given hints when they weren't. They expected
their teacher to either tell them they were doing well, or to ask them

a question. Questions from the teacher were a signal that the student's answer was not correct. The first several weeks of class were always something of a struggle as I worked to change my students' expectations. I thought about the three questions that Schoenfeld (1987) said he reserved the right to ask his students during problem solving. The first question asked the students to describe what they were doing. That was something I already did. The second and third questions were the ones that interested me. These questions asked students to explain why they chose their current approach and how they thought it might help them to find the solution.

In fact, perhaps because my students were so much younger than his, Schoenfeld's questions didn't map specifically onto what my students needed. However, I found I could use his questions to deepen my own thinking about student learning and this enabled me to give the students greater responsibility for their work. I found that the ways I probed when students worked in small groups became more deliberate. In addition to discussing the mathematical content of their problem solving, I would question the groups' choices. Often I would ask the quietest student to tell me why they were solving the problem a particular way. I would also ask whether they thought that this was the best way to do it, and why.

I knew that I was doing something right when a lively, talented girl asked me whether her group's answer was right. She paused, looked at me, and said, "You aren't going to tell us are you?" I must have smiled or nodded. In mock outrage, she cried: "You're going to make us *think*!" (SS, notes)

The content of the articles did not always directly correspond to the classes or students we taught. Like the cylinder problem though, we all had different perspectives on what the articles suggested and how this information applied to practice. Working together to think about research others had conducted led us to clarify our own questions and helped us to understand the potential strengths and needs of our students. For instance, after reading Stigler and Hiebert's (1999) *The Teaching Gap*, the Math Forum BRAP group realized that differences between what we and other teachers ask of our students might constitute a culture shock. We also became aware that in our work together we had been doing informal lesson study all along. We did not use lesson study to describe our practice because *The Teaching Gap* had not yet been published.

Math Forum Support: Collaborating Around
Rich Mathematics and Pedagogy

Collaborating and thinking with others to understand the mathematical and pedagogical possibilities of a rich problem in the Math Forum BRAP project provided us with a basis for working with and making sense of the *Standards* and their implications for our practice. The Math Forum staff members used democratic methods to work with the BRAP teachers to:

- Pose questions and tasks that elicited, engaged, and challenged all participants,
- Listen (or learn to listen) carefully to each person's ideas,
- Clarify and justify ideas orally and in writing,
- Use current discussions of the BRAP group to identify ideas to be pursued in subsequent conversations,
- Decide together when and how to attach mathematical notation and language to ideas,
- Decide when and how to provide information, when to clarify, when to model, when to lead, and when to let a person struggle, and
- Monitor participation in discussions and decided when and how to encourage each other to participate. (Participant observation, BRAP)

The points that had once challenged Judy about how to support discourse about mathematics were lived experiences in the BRAP workshops. The readings, mathematics, online discussions, videotapes of their own and others' classrooms, and ways that the Math Forum staff members facilitated the project, provided models of how discourse about mathematics could be undertaken. This model provided a basis for our continued collaboration and discussion. Our BRAP experiences also developed our view of rich mathematics problems. For instance, as we pushed deeper into the cylinder problem, we discovered that what we originally thought was a rich problem for middle school through, say, geometry level students, turned out to be problem that could challenge our students and all of us into calculus. We found that we sought out non-routine challenge problems because they were rich enough to be visited and re-visited over time and provided all of us (and our students) with opportunities to connect to the problem in one or another way (CTGV, 1997).

We also realized that carrying the question, "How do I know that I'm right?" was a reminder to reflect on and expand our own and our students' repertoires, or toolkits. In this way, we seek evidence for what we know. Then, if we do not have justification, we can figure out what we still need to figure out.

A Collaborative Inquiry: How Do We Support
Students to Think Mathematically?

Sternberg (1985) has noted that the most difficult aspect of problem solving is identifying what the problem is. Presuming that the student (or teacher) does understand what the problem is, the next step is to begin working on a solution strategy. Unfortunately, telling a student the algorithm or providing a teacher with a list of the characteristics of mathematical discourse only points to the type of problem with which the student or teacher needs to work. Having information about what to do is different than the process of beginning to explore and work with this information.

We are very aware that some students can look at an algorithm and seem almost automatically to connect it with the underlying concept. They can identify how it relates to what they have been working on in class previously and they also can point to its real-world application. In fact, many of us as teachers are probably teaching mathematics because we can provide ourselves with this type of information when we look at the problems we present our students. However, many other students and adults have never experienced looking at a mathematics problem and understanding its context and a relevant algorithm (National Research Council, 1989). This does not mean that it is impossible for people to develop an appreciation of an algorithm as more than a rote short cut. It may mean that group collaboration can provide the support that teachers and students need to begin developing an understanding of a problem's context (Blumenfeld et al., 1991; Cobb, 1995; CTGV, 1997; DeCorte, Verschaffel, & Op'T Eynde, 2000). Further, assignments can be structured to enhance the likelihood that students will be able to see alternate possibilities for working with a problem. For example, group work that (a) assigns each group a slightly different form of the question being posed, (b) asks students to explain how they have come to the conclusion that they have, c) asks students as they work on a problem to build a spreadsheet that includes particular dimensions and values, and/or (d) is orchestrated so that the order in which solutions are presented leads the students to consider typical errors, simple single solutions, and finally the generalized case (Stigler & Hiebert, 1999).

As our experiences underscore, support for the development of mathematical thinking involves an ongoing process of assessing learners' strengths and needs. We need to think about the available tools and the learners' readiness to recognize and use them as tools. We must consider the conversations that they have had about mathematics, and the conversations that we would like to help them to have. The goal is to support learners both to ask their own questions and to seek out resources that will help them to find answers—then, by definition, they will also be generating strategies and reflecting on evidence to assess which solutions or strategies are right.

We realize, however, that presenting rich problems to students and using group work in classes may be necessary but not sufficient for supporting students to develop their abilities to think mathematically. Much depends on whether they have developed, or can generate, the set of resources—a toolkit on which we also expect them to draw. In designing tasks for our students, we need to recognize that we are supporting them to develop a repertoire of resources for problem solving, *as well as* the ability to work with a particular concept or problem type. We also need to help them recognize that they have multiple tools, or strategies, available. We need to assure students that it is appropriate to ask questions, especially questions that explain what they have already tried and what they are trying to figure out.

Students who are supported to seek to answer the question, "How do I know that I'm right?" may begin to reflect on the strategies that they do use and those that they could use. Similarly, if prior experience has not provided them with an example of, say, how technology might be used as a tool, then they need support to think of technology as a tool (e.g., thinking of *Geometer's Sketchpad* as a tool for thinking and problem solving generally, rather than as a program you use when you are doing a geometry activity in geometry class). Ideally, such practices provide students with a requirement that they develop and draw on resources in problem solving.

Support to use strategies and reflection in work with mathematics appears ideally to include: (a) other people with whom to think, (b) models of how to approach problems, (c) alternative possibilities for tools to use in problem solution, and (d) recognition that what needs to be accomplished is and will be useful. A succession of experiences that involve use of these resources, positive feelings, a sense of possibility, and rich problems allow students of varying abilities to make connections to mathematical ideas and be challenged to stretch what they understand.

Collaborative inquiry has helped us as teachers to develop, recognize, and use strategies to reflect on the teaching of mathematics. Similarly, collaborative inquiry has supported our students to develop mathematical thinking. As Heaton (2000) notes, however, "Teaching mathematics for understanding is not something that is ever completely learned. One can get better at 'it' but, the teaching, is forever under construction" (p. 141). In the BRAP videopaper, we concluded that the process of talking with colleagues "meant that we documented alternate solutions to similar problems. It ... pushed us to try others' suggestions and to revise plans that we'd thought we knew worked, in turn encouraging us to push our thinking about mathematics teaching" (The Math Forum BRAP Group, 2000).

The present discussion expands on this point. As Shayla's case illustrates, mathematical thinking can benefit from interaction with others, but even if verbal expression may enhance and support its development, it does not always need to be verbal. More precisely, conditions that support reflection involve the interaction of a person and a whole host of potential possibilities and constraints. We as teachers always want to appreciate that each student with whom we work has a slightly different configuration of strengths, needs, experiences, and interest. We want to guard against presuming that we know that we are right as we work. We want to help each other and our students to continue to question what works, as well as how and why it works as it does.

Our findings suggest that supporting mathematical thinking provides teachers *and* students with opportunities to (a) assess what they do and do not understand in a situation, (b) gain experience with alternative solution strategies, and (c) identify tools that allow them to teach and learn. This support includes the content knowledge teachers and students bring to learning, examples and models of potential tools and their usage, as well as encouragement for teachers and students to continue asking and pursuing answers to their questions.

NOTES

1.   The Math Forum's Bridging Research and Practice project was a National Science Foundation sponsored collaboration of TERC and The Math Forum (NSF# 9805289).

2.   *K. Ann Renninger* is a professor of educational studies at Swarthmore College, PA. She teaches a diverse and highly selective student population; she has worked as a teacher, a supervisor of student teachers, and as a researcher in inner city and suburban, public, private, and parochial school settings. She studies interest and learning, and conducts research for The Math Forum (mathforum.org). *Susan Stein* works with middle school teachers and mathematics specialists in the Portland School District (OR), as part of an NSF fellowship through the Center for Learning and Teaching in the West. In addition, she is a doctoral student in Education at Portland State University. The students of the teachers with whom she works come from working class and poor families that are quite diverse culturally and include African American, Native American, European American, Latinos/a, Eastern European, and Asian backgrounds. During work on The Math Forum's BRAP videoproject, Susan was the mathematics department chair and a middle school mathematics teacher at Wilmington Friends School in

Delaware. Students at Wilmington Friends are predominantly European American and middle to upper middle class. *Judith Koenig* is a mathematics consultant who works with middle school teachers in various regions of the country. During work on The Math Forum's BRAP video project, Judy was the mathematics department chair and a middle school math teacher at Nevin Platt Middle School in Boulder, Colorado. Students at Nevin Platt are predominantly Caucasian and middle class. In 1996, Judy was a recipient of the Presidential Award for Secondary Mathematics Teaching. *Art Mabbott* is a Curriculum Consultant/Math coach for several public middle schools in Seattle, WA. The students with whom he works come from middle and lower SES families that represent a wide-range of language, ethnic, and racial backgrounds, including Southeast Asian, Eastern European, Middle Eastern, African, Central & South American and Native American. During work on The Math Forum's BRAP video project, Art was the math teacher in the Prism Program at Odle Middle School and later a member of the math department at Sammamish High School. The students at Odle were participants in a very selective gifted program that served students who were predominantly Caucasian and Asian, and came from middle to upper middle class families. Students at Sammamish are diverse in terms of their socio-economic status, and their language, ethnic and racial backgrounds.

REFERENCES

Atkins, S. (1999). Listening to students. *Teaching Children Mathematics, 5*(5), 289-295.

Ball, D. L. (1993). With an eye on the mathematical horizon: Dilemmas of teaching elementary school mathematics. *Elementary School Journal, 93,* 373-399.

Ball, D. L., & Cohen, D. K. (1999). Developing practice, developing practitioners: Toward a practice-based theory of professional education. In L. Darling-Hammond & G. Sykes (Eds.), *Teaching as the learning profession: Handbook of policy and practice* (pp. 3-32). San Francisco: Jossey-Bass.

Blumenfeld, P, C., Soloway, E., Marx, R. W., Krajcik, J. S., Guzdial, M., & Palincsar, A. S. (1991). Motivating project-based learning: Sustaining the doing, supporting the learning. *Educational Psychologist, 26,* 369-398.

Brown, A. L., Bransford, J. D., Ferrara, R. A., & Campione, J. C. (1993). Learning, remembering and understanding. In J. H. Flavell & E. M. Markman (Eds.), *Cognitive development,* Vol. 3 (pp. 77-166). In P. H. Mussen (Ed.), *Handbook of Child Psychology* (4th ed). New York: Wiley.

Cobb, P. (1995). Mathematical learning and small-group interaction: Four case studies. In P. Cobb & H. Bauersfeld (Eds.), *The emergence of mathematical thinking: Interaction in classroom cultures* (pp. 25-129). Mahwah, NJ: Erlbaum.

Cooney, T. J., & Shealy, B. E. (1997). On understanding the structure of teachers' beliefs and their relationship to change. In E. Fennema & B. S. Nelson (Eds.), *Mathematics teachers in transition* (pp. 87-109). Hillsdale, NJ: Erlbaum.

Cognition and Technology Group at Vanderbilt University (CTGV). (1997). *The Jasper Project: Lessons in curriculum, instruction, assessment, and professional development*. Mahwah, NJ: Lawrence Erlbaum Associates.

Collins, A, Brown, J. S., & Newman, S. E. (1989). Cognitive apprenticeship: Teaching the crafts of reading, writing, and mathematics. In L. Resnick (Ed.), *Knowing, learning, and instruction: Essays in honor of Robert Glaser* (pp. 455-94). Hillsdale, NJ: Lawrence Erlbaum Associates.

DeCorte, E., Verschaffel, L., & Op'T Eynde, P. (2000). Self-regulation: A characteristic and a goal of mathematics education." In M. Boekaerts, P. R. Pintrich, & M. Zeidner (Eds.), *Handbook of self-regulation* (pp. 687-726). New York: Academic Press.

Dewey, J. (1933). *How we think*. Buffalo, NY: Prometheus Books.

Eccles, J., Adler, T. F., Futterman, R., Goff, S. B., Kaczala, C. M., Meece, J. L., et al. (1983). Expectancies, values, and academic behaviors. In J. T. Spence (Ed.), *Achievement and achievement motives* (pp. 75-146). San Francisco: W.H. Freeman and Co.

Gee, J. P. (1999). *An introduction to discourse analysis: Theory and method*. London: Routledge.

Goldsmith, L., & Schifter, D. (1997). Understanding teachers in transition: Characteristics of a model for developing teachers. In E. Fennema & B. S. Nelson (Eds.), *Mathematics teachers in transition* (pp. 19-54). Hillsdale, NJ: Lawrence Erlbaum Associates.

Heaton, R. M. (2000). *Teaching mathematics to the new standards: Relearning the dance*. New York: Teachers College Press.

Lampert, M. (1986). Knowing, doing, and teaching multiplication. *Cognition and Instruction, 3,* 305-342.

Lampert, M. (1998). Investigating teaching practice. In M. Lampert & M. L. Blunk (Eds.), *Talking mathematics in school: Studies of teaching and learning* (pp. 153-162). New York: Cambridge University Press.

Lampert, M. (1990) When the problem is not the question and the solution is not the answer: Mathematical knowing and teaching. *American Educational Research Journal, 27,* 29-63.

Lappan, G., Fey, J. T., Friel, S., Fitzgerald, W. M., & Phillips, E. D. (1998). *Comparing and scaling: Ratio, proportion, and percent. Connected Mathematics Project*. Menlo Park, CA: Dale Seymour Publications.

Mandler, G. (1989). Affect and learning: Causes and consequences of emotional interacting. In D. B. McLeod & V. M. Adams (Eds.), *Affect and mathematical problem solving* (pp. 3-19). New York: Springer-Verlag.

Markus, H., & Nurius, P. (1987). Possible selves. *American Psychologist, 4,* 954-69.

Marx, R. W., Blumenfeld, P. C., Krajcik, J. S., & Soloway, E. (1998). New technologies for teacher professional development. *Teacher and Teacher Education, 14,* 33-52.

The Math Forum BRAP Project. (2000). *Encouraging mathematical thinking: Discourse around a rich problem*. Retrieved on the World Wide Web: mathforum.org/brap/wrap.

National Council of Teachers of Mathematics. (1991). *Principles and standards for school mathematics*. Reston, VA: Author.

National Council of Teachers of Mathematics. (2000). *Principles and standards for school mathematics*. Retrieved on the World Wide Web: http://standards.nctm.org/document/index.htm.

National Research Council (1989). *Everybody counts: A report to the nation on the future of mathematics education*. Washington, DC: National Academy Press.

Noddings, N. (1992). *The challenge to care in schools: An alternative approach to education*. New York: Teachers College Press.

Renninger, K. A. (2000). Individual interest and its implications for understanding intrinsic motivation. In C. Samsone & J. M. Harackiewicz (Eds.), *Intrinsic and extrinsic motivation: The search for optimal motivation and performance* (pp. 375-407). New York: Academic Press.

Renninger, K. A, Farra, L., & Feldman-Riordan. C. (2000). The impact of The Math Forum's problems of the week on students' mathematical thinking. *Proceedings of ICLS 2000*. Mahwah, NJ: Lawrence Erlbaum Associates. Retrieved from the World Wide Web: www.mathforum.org/articles/rennin2_2000.html.

Renninger, K. A., Sansone, C., & Smith, J. L. (2004). Love of learning. In C. Peterson & M. E. Seligman (Eds.), *Character strengths and virtues: A classification and handbook*. (pp. 161-179) New York: Oxford University Press.

Reinhart, S. C. (2000). Never say anything a kid can say! *Mathematics Teaching in the Middle School, 5*, 478-83.

Resnick, L. (1988). Treating mathematics as an ill-structured discipline. In R. I. Charles & E. A. Silver (Eds.), *The teaching and assessing of mathematical problem solving* (pp. 32-60). Hillsdale, NJ: Lawrence Erlbaum Associates.

Rodgers, C. (2002). Defining reflection: Another look at John Dewey and reflective thinking. *Teachers College Record, 4*, 842-866.

Rothschild, K. J. (1997). *Dancing at two weddings: A case study of mathematics reform.* (Unpublished doctoral dissertation, University of Pennsylvania).

Sansone, C., & Berg, C. A. (1993) Adapting to the environment across the life span: Different process or different inputs? *International Journal of Behavioral Development, 16*, 215-41.

Schoenfeld, A. H. (1987). What's all the fuss about metacognition? In A. Schoenfeld (Ed.), *Cognitive science and mathematics education* (pp. 189-215). Hillsdale, NJ: Lawrence Erlbaum Associates.

Schoenfeld, A. H. (1991). On mathematics as sense making: An informal attack on the unfortunate divorce of formal and informal mathematics. In D. Perkins, J. Segal, & J. Voss (Eds.), *Informal Reasoning in Education* (pp. 311-42). Hillsdale, NJ: Lawrence Erlbaum Associates.

Schoenfeld, A. H. (1992). Learning to think mathematically: Problem solving, metacognition, and sense making in mathematics. In D. A. Grouws (Ed.), *Handbook of research on mathematics teaching and learning: A project of the National Council of Teachers of Mathematics* (pp. 334-70). New York: MacMillan Publishing Co.

Sternberg, R. J. (1985). *Beyond IQ.* New York: Cambridge University Press.

Stigler, J. W., & Hiebert, J. (1999). *The teaching gap.* New York: Free Press.

Webb, N. L., & Palincsar, A. S. (1996). Group processes in the classroom. In D. C. Berliner & R. C. Calfee (Eds.), *Handbook of educational psychology* (pp. 841-873).

Wigfield, A., & Eccles, J. S. (2002). The development of competence beliefs, expectancies for success, and achievement values from childhood through adolescence. In A. Wigfield & J. S. Eccles (Eds.), *The development of achievement motivation* (pp. 92-122). San Diego, CA: Academic Press.

Zimmerman, B. J., & Risemberg, R. (1997). Self-regulatory dimensions of academic learning and motivation. In G. D. Phye (Ed.), *Handbook of Academic Learning: Construction of Knowledge* (pp. 105-25). San Diego, CA: Academic Press.

CHAPTER 8

MEANINGFUL MATHEMATICS FOR URBAN LEARNERS

Perspective From a Teacher-Researcher Collaboration

Christine D. Thomas
Georgia State University, Atlanta, GA

Carmelita Santiago
Atlanta Public Schools, Atlanta, GA

In our quest to engage urban seventh-grade students in meaningful experiences in the mathematics classroom, our partnership evolved as a teacher-researcher collaboration. We, Carmelita, a seventh-grade mathematics teacher and Christine, a university mathematics education researcher, focused our efforts on engaging Carmelita's students in high-quality mathematics lessons. Our purpose was to apply our collaborative work in advancing urban learners through mathematics. With this in mind, we formulated a question that would guide our work as well as our research. In what ways can a teacher-researcher collaboration make mathematics meaningful for

Teachers Engaged in Research
Inquiry Into Mathematics Classrooms, Grades 6–8, pages 147–170

students in an urban school? Our question kept us focused and engaged in a successful and notable collaboration. This is our story.

ESTABLISHING A
TEACHER-RESEARCHER COLLABORATION

The stimulus for our partnership grew out of our relationship as partners in reflective practice through implementation of the Reflective Teaching Model (RTM). The RTM is a teacher-change model developed in the Atlanta Math Project (AMP) at Georgia State University (Hart, 2002; Hart & Najee-ullah, 1997; Hart, Schultz, & Najee-ullah, 1992). It is an innovative process of reflective lesson planning, teaching, and debriefing designed to facilitate standards-based teaching as documented in the National Council of Teachers of Mathematics (NCTM) Standards (1989, 1991, 1995, 2000). The RTM goals are:

- to provide teachers with experiences that foster the construction of new knowledge about the teaching and learning of mathematics.
- to promote the implementation of standards-based instruction.
- to study how teachers' knowledge and classroom environments change over time.

Teacher participation in the RTM begins with initial professional development experiences that introduce them to the spirit and philosophy of the RTM. This initial experience is followed by academic-year experiences in their classrooms with subsequent professional development. All participating teachers are paired with a mentor who serves as a nurturing partner. The major role of the mentor is to serve as partner in the change process while assuming a nonjudgmental stance and posing questions that stimulate the teacher's ability to reflect critically on her practice (Keys & Golley, 1996).

All activities of the RTM are set in a *model/experience/reflect* framework (Hart, Najee-ullah, & Schultz, 2004). Each component comprises a process in itself. The RTM professional development incorporates *models* that provide teachers the opportunity to see standards-based teaching practices in action. While studying standards-based instruction in action, participating teachers are simultaneously engaged in RTM *experiences*. These experiences are based in (a) implementing inquiry-based lessons, (b) deepening teachers' understanding of mathematics, (c) developing teachers' confidence in implementing reformed teaching strategies, and (d) advancing teachers' abilities to critically *reflect* on practice.

During the classroom follow-up phase of the RTM, the teacher and mentor participate in *plan-teach-debrief* cycles. The *planning* session encourages the teachers to think about student approaches to mathematics and instructional strategies that promote student engagement and facilitate students' abilities to construct knowledge.   While *teaching* the lessons, teachers become adept in responding to unanticipated student questions and facilitating discourse. The *debriefing* sessions are non-evaluative sessions in which the teachers are expected to analyze their teaching in light of what they planned and how they taught the lesson.

As partners in this process, we began our relationship as teacher and mentor. Our relationship has evolved over time as we engaged continuously in implementing RTM practice. Through this process, we have grown in our perspective of the dynamics of, who we are, what we choose to do, and how we go about our work. Our perspective on our partnership has grown from a teacher-mentor relationship to partners in research.

OUR INDEPTH STUDY OF CARMELITA'S CLASSROOM

As research partners, we decided to engage in a long-term study on the impact of standards-based teaching in Carmelita's classroom. We chose to collect data during a time when Carmelita would implement an innovative unit through which she would teach all of the seventh-grade geometry objectives over a twenty-day period. After several discussions about the approach that would be implemented to teach the school system's seventh-grade geometry objectives in her classroom, Carmelita developed a project-based unit. As the geometry unit was developed, we simultaneously designed the research study that would be conducted in conjunction with the teaching of the unit.

The context for our work was an inner city middle school with approximately 850 students in grades six through eight. The school is located in an urban, predominantly African American community in the southeast. The student population is approximately 99% African American and 1% Hispanic with approximately 78% of the students on free and reduced lunch. A majority of the students live in low-to-middle level income homes with a single parent and are latchkey children. Carmelita's seventh-grade mathematics class was comprised of 18 females and 12 males between the ages of 11 and 13. The students' ability levels were mixed heterogeneously from average to above average.

Within the school there was a constant effort to overcome the struggles and challenges associated with schools that serve low-income communities where over 50% of the student population qualifies for free or reduced lunch (Kitchen & DePree, 2005). These schools "have unique sets of problems that distinguish them from their more affluent, suburban counterparts" (p. 3). In our setting, a major school-wide challenge was to improve student performance in mathematics. For all grade levels, student performance in mathematics fell below the cutoff rate on the state's standardized test. Finding ways to address this challenge gave meaning to our work as a teacher-researcher collaborative in an urban setting.   Therefore, as we approached the development and implementation of the project-based unit as well as our research study, we remained focused on our agenda to advance the mathematical learning of our urban students.

OUR APPROACH TO THE RESEARCH STUDY

We collected data over the twenty-day, project-based unit implementation. All classes were videotaped. Other sources of data included student work, field notes collected during class observations, Carmelita's journal, and weekly audio taped debriefing sessions. We focused on our guiding question for our work—In what ways can a teacher-researcher collaboration make mathematics meaningful for students in an urban school? In an effort to further define our study, we added a second question—In what ways does the Reflective Teaching Model facilitate high performance mathematics teaching for urban African American learners?

We used the RTM plan-teach-debrief cycle to guide our research process. While engaged in the debriefing sessions, we applied an indepth approach to the analysis of the planning and teaching. In analyzing the planning and teaching, we applied a questioning and probing technique developed for RTM debriefing. Samples of debriefing questions are:

- How do you think the lesson went?
- Were you able to implement the lesson as planned? In what ways did the planning and teaching remain the same and how did they differ if at all?
- Did the student learn what you intended? If yes, how do you know? If no, how do you know?
- What do you think did not work well in the lesson?
- What do you think worked well during the lesson?
- What would you do differently?
- What would you keep the same?

We analyzed the video tapes and examined student work. As we grew in our level of analysis, we began to focus our attention on additional literature to inform our work. We examined the literature on culturally relevant pedagogy (Ladson-Billings, 1995a, 1995b) and the Mathematics Teaching Cycle (Simon 1995, 1997). This focus added another layer to the depth of our data analysis and provided directions for classroom interventions as the unit developed. Culturally relevant pedagogy rests on three criteria:

(1) students' experiences in academic success, (2) students' development and maintenance of cultural competence, and (3) students' development of a critical consciousness through which they challenge the current social order (Ladson-Billings, 1995). Application of the Mathematical Teaching Cycle (MTC) in our analysis showed us how we were aligned with this framework in teaching. "According to the MTC, the mathematics teacher's actions are at all times guided by his or her current goals for student learning, which are continually being modified based on interactions with students" (Simon, 1997, p. 77).

As we began to integrate the literature explicitly into our practice, the complexity of our work became overwhelming. Therefore, we began to think of ways in which we could bring coherence to our complex work and felt compelled to develop a graphic showing the dynamic interactions among all the components (see Figure 8.1).

As shown, we viewed the RTM as the rudimentary component of all that we do. Following the graphic from the RTM downward to the left and to the right simultaneously, the first level of RTM influence was in our work with teacher narratives in our teacher-researcher collaboration. Given the fact that our responses to the debriefing questions were given in narrative form, we determined this use of teacher narrative to be a key component of our analysis. All of our debriefing sessions were audio taped, transcribed and

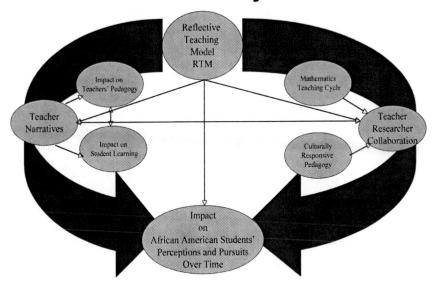

Figure 8.1.   Graphic illustrating data analysis framework.

later analyzed. As we examined our narratives, we were able to pinpoint components of culturally relevant pedagogy that were present during the implementation of the project-based unit. Continuing along Figure 8.1, the arrows indicate the many paths of influence among the components, from the RTM, to teacher narratives, to the teacher-researcher collaboration, then ultimately to the impact on students' perceptions, pursuits, and performance in mathematics. In view of the fact that the research is aligned with the implementation of the project-based unit, it is pertinent that we provide details about the design and implementation of the project-based unit as well as a closer view of student engagement.

A PROJECT-BASED UNIT:
GEOMETRY AND THE ICOSAHEDRON

The geometry unit titled "From Point to Polyhedron: The Icosahedron," was designed as an inquiry-based project through which students could study the basic concepts of geometry from one to three dimensions while actively engaged in constructing icosahedra (Thomas & Santiago, 2002). Standards-based strategies were incorporated in ways to challenge and advance the mathematical performance of the students. The five process standards (NCTM, 2000) were emphasized throughout the unit. Particular attention was focused on the four areas of expectations delineated in the Grades 6–8 Standard on *Geometry* (NCTM, 2000). The unit was designed to engage students in:

- analyzing characteristics and properties of two- and three-dimensional geometric shapes and developing mathematical arguments about geometric relationships;
- specifying locations and describing spatial relationships using coordinate geometry and other representational systems;
- applying transformations and using symmetry to analyze mathematical situations; and
- using visualization, spatial reasoning, and geometric modeling to solve problems. (NCTM, 2000, p. 232)

Each step in building icosahedra represented the study of specific concepts in geometry. The unit was opened from a synthetic approach to geometry where the basic units of geometry, point and line were intuitively developed. The students began by developing a concept for why the terms are called undefined terms in geometry. Next the undefined terms, point and line were used to define terms such as segment, ray and angle. Through this synthetic approach the concept of point was introduced as an exact location in space, having no size or shape. This led to the discussion that two points determine exactly one line. Next, rays and segments were investigated as parts of a line and segments as part of a ray. After exploring these basic terms of geometry, the construction of the icosahedra began. The icosahedra were built with drinking straws and yarn. Drinking straws were used to represent segments. The segments (drinking straws) were connected edge to edge by inserting and then knotting yarn inserted through the tubes of the straws.

Investigations in two dimensions began with two straws connected end to end representing rays to form angles. The students identified one ray as the initial side, then rotated the second ray as the terminal side through various degrees from zero to 360. They investigated and identified angles as acute, right, obtuse, straight, and reflex.

After completing the lessons on angles, the students connected the third segment to construct the first triangle, and began the study of polygons. Although only equilateral triangles were produce when connecting the straws, investigations were extended to all types of triangles by angle measures and by side lengths. Students explored various types of triangles by building models and measuring angles and sides. They also used *Geometer's Sketchpad* to investigate triangles. They discovered that the sum of the angles of a triangle is 180 degrees, as well as other properties. Vocabulary terms that evolved as a result of the investigations included triangles as acute, right, obtuse, equiangular, scalene, equilateral or isosceles.

The next step was to build the midsections of the icosahedra by connecting a string of adjacent triangles. This began with a discussion of how many segments were needed to add a second triangle. The students added two segments to build the second triangle; then continued to add segments to build a string of adjacent triangles. Figure 8.2 shows students holding up a

Figure 8.2.   Students holding strings of adjacent triangles.

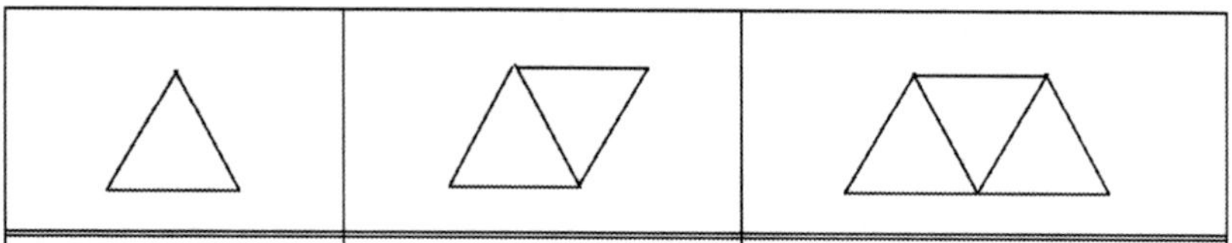

Triangle	Parallelogram	Trapezoid
1- Triangle	2 - Triangles	3 - Triangles
3 - Straws	5 – Straws	7 - Straws

Figure 8.3.   Polygon moving from triangle to trapezoid
with additions of triangles.

string of adjacent triangles. As additional triangles were added to the original triangle, the overall model changed from a triangle to a parallelogram to a trapezoid. After the first trapezoid, with each additional triangle, the overall model alternated from a parallelogram to a trapezoid (see Figure 8.3).

At this stage, an indepth exploration of quadrilaterals and the relationship among them was developed in the unit. While studying the trapezoid, the students began an investigation of parallel lines, transversals, and the angles formed when two parallel lines are intersected by a transversal. In Figure 8.4, Carmelita is engaging the students in exploring the trapezoid.

Next, connecting the string of ten adjacent triangles end to end completed the midsections of the icosahedra. Figure 8.5 shows students with the midsections of icosahedra on their desks. Next, the students completed the domes for the top and bottom to complete the icosahedra. With the completed Icosahedra, we began the study of three-dimensional geometric figures. Students were then engaged in the study of polyhedra. We studied topics at this stage of volume, surface area, Euler's formula for vertices, faces and edges, and the Platonic solids.

The Process Standards

We examined and applied processes intricately in all lessons as we focused on strategies that would foster students' abilities to construct mathematical understanding. We developed lessons from a problem-solving perspective and designed them to engage students in making connections within

Figure 8.4.   Carmelita leading students in exploring the trapezoid.

Figure 8.5.   Students with the midsections of
icosahedra The Process Standards.

mathematics and to other disciplines. The lessons were also designed to stimulate and encourage communication of ideas and concepts from a variety of perspectives including through the use of representations and reasoning and proof. The developing icosahedra provided a plethora of representations for figures in geometry from one to three dimensions. Reasoning and proof were apparent as students confirmed their understanding of the mathematical tasks. For example, when using *Geometer's Sketchpad*, students were involved in modeling, conjecturing, inductive reasoning, making arguments and validating claims. All processes were inextricably linked within the lessons. In the following sections, we examine problem solving, connections and communications within the context of the geometry unit.

Problem solving. The entire unit was developed as an inquiry-based project. Lesson planning included the construction of non-routine problems within the context of the daily objectives. Designing effective questions was the major approach to problem solving. Opening questions and follow-up questions were developed. After engaging students in a lesson through the opening question, follow-up questions were used to promote student thinking and to address questions that students might ask during problem solving. Let's examine one illustration of how lessons were approached through problem solving.

The series of lessons on classifying triangles advanced along three levels. In level one, the students were expected to find all types of triangles by angle measures (acute, right, obtuse) and side lengths (scalene, isosceles, equilateral). Once the students found six types of triangles, they were challenged to prove that there were only six types. The problem posed was, "How can we be sure there are no other possibilities?"

In level two, students were asked to determine whether or not a triangle could be classified as more than one type. Questions used to promote investigations in level two included: (a) Can a triangle be both scalene and right?, (b) Can a triangle be isosceles and acute?, (c) Can a triangle be equilateral and obtuse?, and (d) Can a triangle be isosceles and right? In moving towards level three, we expected that students would begin to write their own questions and conduct investigations for solutions. Some of the questions developed by students during the lessons on classification of triangle were: "If a triangle can be classified as more than one type such as isosceles and right, tell why it can be both." and "If a triangle is classified as scalene, what other classifications can that triangle have?" Finally in level three, we designed the lesson to challenge the students to develop a chart showing the relationships among the types of triangles. We anticipated that as more complex problems were introduced, students would become more adept with problem-solving strategies and naturally inclined to pursue mathematical tasks.

Connections. "If students are to become mathematically powerful, they must be flexible enough to approach situations in a variety of ways and recognize the relationships among different points of view" (NCTM, 1989, p. 84). This project-based geometry unit reflected this spirit of developing mathematically powerful students through connections. Connections were made within geometry, to measurement, to algebra, to discrete mathematics and to other disciplines. Within geometry connections of the icosahedra were made to the Platonic Solids, Fibonacci Sequence, Golden Rectangle, Golden Ratio, and Euler's formula for vertices, edges and faces (Fetter, Schmalzried, Eckert, Schattschneider, & Klotz, 1991; Pappas 1989). Specific connections were made to science when students conducted research on the Platonic Solids. Figure 8.6 shows one student inserting his golden rectangles in the interior of his icosahedron. Through internet and library research, students learned how crystals grow in the shape of polyhedra and about Plato and the Platonic solids.

Connections were made to algebra from a variety of perspectives. Algebraic thinking was infused as the students built the midsections of the icosahedra as a string of adjacent triangles. Students were asked, "Can we determine an algebraic equation that will tell us how many straws are needed to build a string with any given number of triangles?" In this lesson, the students looked for patterns by investigating a string with one triangle,

Figure 8.6.   Student inserting his golden rectangles
in the interior of his icosahedron

next two triangles, then three triangles, and so on (see Table 8.1). Next, the students made a table of values, then derived an algebraic equation for the table of values.

We also introduced discrete mathematics. While the developing icosahedra were in the two-dimensional stage of adjacent triangles, we deemed the study of networks an appropriate connection. The string of adjacent triangles served as a natural model for studying networks consisting of vertices and arcs (edges). This study included investigations of Euler paths, Hamilton circuits and the Königsberg Bridge Problem (Pappas, 1989). After working with a stripped version of the picture of the bridges, which served as a graph of the problem, and discussing the terms vertices and edges with respect to a network, the students worked with some basic graphs using Euler's definitions of paths and circuits. These lessons led to internet and library research on graph theory and on the mathematicians Euler and Hamilton.

Communication. Establishing a classroom climate and culture in which urban students would be comfortable in expressing their ideas as well as skillful in listening to their peers express their ideas was a focal point in implementation of the unit. We wanted to establish a non-threatening classroom climate which would be conducive for student engagement in ways that promoted development of confidence in their mathematical abilities. This meant establishing effective means of communication from a variety of perspectives. As such, prior to beginning the unit, a class period was devoted to developing listening skills, time management skills and social skills for working in cooperative groups. Strategies to enhance and maintain these skills were integrated daily.

Table 8.1.   Students used this table to find an equation relating these values

Number of Triangles	Number of Segments
1	3
2	5
3	7
•	•
•	•
•	•
n	?

Writing in mathematics was one specific strategy that we placed prominently within the unit. Each student was responsible for developing and organizing a project notebook and provided instructions on how to organize the research notebook. The notebook was organized with sections for vocabulary, daily reflections, essays, internet research, library research, handouts, assessments, lab notes and a daily log of the developing model. Students individualized their notebooks by designing front and back covers. The front cover included a title, the student's name, and the teacher's name; the students created these covers using computers. The back cover was titled "About the Researcher." On the back cover the students inserted their pictures taken with a digital camera and inserted their narratives about themselves as researchers. Students were expected to record in their lab logs the outcomes of what they observed through building the model each day. At the end of each day's investigation and recording of findings in the log, students wrote a reflection summarizing what they learned.

Inside the Classroom

To begin the unit, students were given a list of materials to bring that included a three ring binder with plastic front and back cover slips, notebook paper, a ruler, protractor, compass, 100 straws, a specific color skein of yarn, and a pair of scissors. Students were assigned to research teams of varying abilities, genders and personalities. Each research team was identified with a specific color of yarn. Research team colors were green, blue, red, yellow, and orange. Students were given written job descriptions of their responsibilities to their research teams.

Once the unit was in progress, the overall classroom climate evolved as a laboratory of research teams intensely engaged in the study of mathematics through this project-based approach. Given our knowledge of the NCTM *Standards* (1989), recommendations on student disposition toward mathematics and our knowledge of urban students learning styles (Walker & Chappell, 1997), we incorporated strategies that would foster our students' dispositions towards mathematics as well as their engagement in the cognitive concepts. NCTM's recommendation on disposition toward mathematics is as follows:

Learning mathematics extends beyond learning concepts, procedures, and their applications. It also includes developing a disposition toward mathematics and seeing mathematics as a powerful way for looking at situations. *Disposition* refers not simply to attitudes but to a tendency to think and to act in positive ways. Students' mathematical dispositions are manifested in the way they approach tasks—whether with confidence, willingness to explore alternatives, perseverance, and interest—and in their tendency to reflect on their own thinking. (p. 233)

Some of the strategies applied in fostering students' dispositions included establishing cooperative groups in which students supported each other in learning, and engaging students in a non-threatening classroom climate where they became accustomed to presenting ideas that may not always be correct. Through observation, we assessed each student's development of a personal meaning for the value of engaging in and pursuing mathematical tasks. It was apparent that students' dispositions toward mathematics had been impacted as members of this community of researchers. Excerpts from students' daily reflections as shown below provide testimony to the impact on students' dispositions and pursuits of mathematical tasks.

Student 1: Day 3

Today, I'm ready to learn more than I learned yesterday. I realize that today is Wednesday and another school day but I seem like I'm ready. I really understood parallel lines and transversal lines. We made a trapezoid and labeled the parallel lines and all transversal lines. I feel like I have done a lot. I can't wait until tomorrow.

Student 2: Day 5

Today is Friday, November 5th and we are looking at a video. The video showed a lot of things that I didn't know about the Platonic Solids. I learned that there are five different Platonic Solids and they are Icosahedron, Tetrahedron, Cube, Octahedron, and Dodecahedron. Each Platonic Solid is different from the other. I also learned that fire is Tetrahedron, air is Octahedron, water is Icosahedron, earth is Cube, and the universe is Dodecahedron. Platonic solids have different examples like crystals, organisms and molecules. I also learned the dual of Platonic Solids. Today I felt like I learned a lot. I can't wait until Monday.

Student 3: Day 12

Today is Wednesday, November 10th, and I am in a great mood. When we first got to class, we talked about the golden rectangle. A few students had to leave the class, but when they got back, they were showing the rest of the class how to construct the golden rectangle. When they finished, I helped a team member who did not understand. I showed him how to find the length of his rectangle by multiplying the width by 1.618. Tonight I will construct my golden rectangle and put the bottom on my icosahedron.

WHAT WE LEARNED

When reflecting on all aspects of our work, we were able to examine our learning with respect to the six principles for school mathematics—curriculum, learning, teaching, equity, technology, and assessment (NCTM, 2000). Although we did not begin with the six principles as a framework for analyzing our work, we recognized the ways in which the principles were immersed in our work through data analysis. Throughout our analysis, we found examples of our approaches to curriculum, learning, teaching, equity, technology and assessment. Our vision for establishing and sustaining high-quality and engaging mathematics instruction for urban learners was confirmed through data analysis that revealed an alignment with NCTM's (2000) vision as described in the *Principles and Standards for School Mathematics*.

As we examined our data, we found ways in which components of culturally relevant pedagogy, the Mathematical Teaching Cycle, and the *Principles and Standards* were concentrated in our work and the ways in which they were pertinent to our research question—In what ways can a teacher-researcher collaboration make mathematics meaningful for students in an urban school? We found specific examples of how RTM practice contributed to our ability to make mathematics meaningful for urban learners. We determined through our analysis that making mathematics meaningful was grounded in our continuation of RTM practice based in the implementation of the *Standards*. Our collaboration was rich and meaningful because the *Standards* were not something that we talked about, but were an integral part of the teaching and learning in Carmelita's classroom. Whether or not we discussed and examined the *Standards* explicitly or implicitly in our debriefing sessions, analysis of the data showed the *Standards* were always driving the way in which mathematics was taught and the way in which the students engaged in learning. While we focused our attention on the process standards when planning lessons, in teaching and in the debriefing, we discovered in our analysis of the narratives that the *Principles* were always present in our discussions. Through data analysis, it became apparent that our teacher-researcher collaborative was meaningful and rich, and grounded in our commitment to the implementation of a standards-based approach to teaching and learning mathematics. Earlier in the chapter we described how the process standards were incorporated in lesson planning and implementation. Next we share what we found with respect to the NCTM *Principles* in our analysis of the data.

Curriculum. In the spirit and philosophy of the *Principles and Standards*, the geometry unit was developed as a challenging curriculum. The unit was rich, coherent, and innovative with the process standards embedded throughout the unit. The excitement and power of a challenging curriculum for these urban students was evinced in their pursuits of the mathematical tasks and their dispositions toward learning. Through implementation of this project-based based unit, we learned how to advance the mathematical learning of urban students.

Learning. Since learning was approached from a problem-solving perspective, it was important to give students time to assimilate their understanding at the end of each class period. This was accomplished by giving students time to reflect on their learning and to record their daily reflections. An examination of the students' written reflections showed the impact of reflection on student learning. Through critical reflections, students were expected to examine what they learned and how they learned it. In our analysis of students' reflections, we learned how essential daily reflection was to the development of autonomous learners who were confident in their mathematical knowledge and ability. When we examined the students' journals, we found a consistent pattern across students' journals. This pattern showed continuous vocabulary development in their writing. We found examples of the use of their internet research in their explorations and discussions of topics such as the Golden Ratio, the Golden Rectangle, the Platonic Solids, Fibonacci Sequence and other topics. Examples of advanced mathematical thinking were prominent throughout the students' writing.

However, the ultimate proof of student learning for school level and district level administrators is in how well students perform on standardized tests. Throughout the school year, teachers are compelled to work with their students in ways that maximize student performance on standardized tests. What this means in many cases is "teaching for the test." While the pressure of focusing on standardized testing was not driving our approach to the curriculum and instruction, it is important to note Carmelita's students were not placed at a disadvantage in test preparation. Over time, it became apparent that a standards-based classroom can be an effective method for increasing student performance in mathematics on standardized tests. On the state's standardized test given in the spring of the year, Carmelita's seventh-grade students outperformed all other seventh-grade students in the school.

Teaching. RTM practice was our approach to teaching and teacher change. Our focus in all of the debriefing sessions began with the question, "How do you think the lesson went?" In each debriefing session, we examined how the lesson opened and flowed with respect to teaching. In one of the sessions, Carmelita's first thoughts were: "I am a student of my own teaching. I was never taught mathematics the way in which I teach my students. This is so exciting."

After this debriefing session, our interest in exploring other ways to examine teaching for urban learners escalated. Therefore, we decided to explore ways in which we could explicitly use the components of culturally relevant teaching and the Mathematics Teaching Cycle as lenses for examining our teaching. We applied these theoretical perspectives in our practice. We used the tenets of culturally relevant pedagogy to examine our approach for (a) engaging students in academic success, (b) facilitating student development of cultural competence, and (c) engaging urban learners in a critical consciousness that moved them beyond seeing themselves as inept in mathematics. Then we applied the Mathematics Teaching Cycle in examining the relationships among teacher knowledge, goals for students, anticipation of student learning, planning and our interaction with students.

Equity. The focus on equity in this classroom was centered on gender, learning styles, and socio-economic status. Students were placed in cooperative groups to maximize their engagement in the classroom. Effective implementation of cooperative group learning was practiced through RTM professional development. Students were deliberately assigned to teams in ways that each individual's performance could be enhanced. Careful attention was given to the impact of cooperative group work on the performance of African American males and African American females. Through the use of cooperative groups, students were afforded a plethora of opportunities to engage in meaningful discussion about their ideas in mathematics.

With respect to gender equity, we examined how the classroom climate supported males and females in their pursuits of mathematics. We found the project-based approach to learning to be a valuable vehicle for both males and females. Carmelita was pleased particularly with the persistence of the females in building the icosahedra. Carmelita hypothesized at the onset of the unit that an all male team would far exceed an all female team in every aspect because she believed that the males would have greater motivation for building the models than females. In fact, she developed an instructional plan with the idea that she would have to really encourage the girls to participate. Carmelita believed that the girls might not achieve without boys in their groups to challenge them. However, her hypothesis was totally incorrect. As the days progressed, Carmelita found the girls to be competitive and totally engaged in developing the model.

Special attention was given to learning styles in establishing an equitable classroom in which all students were afforded the optimal opportunity to learn. While noting that a curriculum for urban students should incorporate the same mathematical content as the curriculum for suburban and rural students, we were also aware of the research on learning styles and the need to contextualize the mathematics through instructional strategies designed to engage urban learners (Walker & Chappell, 1997). We were conscientious in incorporating the modalities; we took further steps to incorporate learning styles from other perspectives. One of the major approaches emphasized was Howard Gardner's work in multiple intelligences (Gardner, 1993, 1994).

Access to materials and resources is always an issue when teaching students from low-income families. Special attention was given to students' access to the required materials and resources when planning the unit. We knew that all students might not be able to purchase the materials needed for the unit, so we contacted local businesses for donations. Through these donations, we received an abundance of materials and resources and these were available for all students to use throughout the project.

Technology. The use of calculators and computers were integral to student engagement and student learning. Students were allowed to use calculators as needed. *Geometer's Sketchpad* was used whenever appropriate and used often. Additionally, the students used *Microsoft Word, PowerPoint,* and *Excel.* The internet was used for research. Through the use of technology, we witnessed the power of these tools in promoting student learning. In particular, we learned the power of technology in facilitating urban learners through advanced mathematical concepts.

Assessment. Fundamental to every lesson was the attention to assessment from multiple perspectives. Assessment of the curriculum, teaching, learning, and equity were embedded through our consistent RTM practice. As we reflected on our approach to assessment, it became apparent that assessment was "a routine part of the ongoing classroom activity rather than an interruption" (NCTM, 2000, p. 23).

CARMELITA'S VIEWS ON DEVELOPING AS A RESEARCHER

Exploring one's practice by opening up one's self to exploration by others was not an easy task. Reflective practice required that I plan a lesson, share the ideas for implementing my lesson with a colleague, videotape the lesson, share the videotape with the same colleague, and finally discuss what went well and what I would change through a reflective discussion. Christine and I never discussed the idea that this process was a situation of self-imposed case study research. As I think about the process, though, I believe it was. I felt very much a novice and did not understand the concept of conducting a rigorous research study. I remember feeling a bit uncomfortable being in this position as I agreed to participate in the process. However, as I continued my participation I became more and more aware of the power of reflective practice, particularly with respect to the RTM. What I realized was that I absolutely enjoyed how the experiences led me to engage my students in higher-level mathematical thinking. Two essential components of the RTM that fostered my approach to pedagogical reform were my *engagement in paired problem solving* and the *plan-teach-debrief cycles.* While participating in paired problem-solving activities, I developed skills in facilitating a person's thinking through problem solving by asking probing questions. Through my immersion in the plan-teach-debrief cycles, I learned to approach all lessons from a problem-solving perspective and to reflect upon and analyze my teaching during each phase of the cycle. These experiences, coupled with other RTM activities, enhanced my skills in: (a) listening to students as they talked through their understanding in mathematics, (b) providing adequate wait time, and (c) paying attention to and nurturing learning styles. In the process I became more skillful in following up with probing questions based on what I heard the students say and not with my own thinking about the mathematics.

I remember thinking that I needed to give my urban students every advantage and to somehow make a concerted effort to prove that presumed disadvantages did not preclude their ability to understand abstract concepts, even though they were in middle school. I understand now that it was at this point in my career that I became consciously a researcher as a classroom teacher. Reflecting on one's practice is one way to begin the process of understanding how a teacher is a researcher as you consciously look at what you decide to teach and how your teaching impacts student learning.

CHRISTINE'S VIEWS ON THE PARTNERSHIP

What I believe to be a significant influence of my perspective on and engagement in our partnership are my experiences and lessons learned through my personal, professional development as a standard-based practitioner. So, I will begin there with my story. My pathway to standards-based teaching commenced while I was a high school mathematics teacher. During my tenure as a high school mathematics teacher, I accepted an invitation to participate in the Atlanta Math Project (AMP) and as such was immersed in all components of the RTM from the initial professional development through classroom implementation of the plan-teach-debrief cycle and follow-up professional development. I remember how my teaching changed from a "talk and chalk" approach to a student-centered approach. Over time, I developed confidence in my skills in reformed teaching. I witnessed the impact of my change in my students' actions and approaches to learning mathematics and an increase in student achievement. After two years of participating in AMP, I became a mentor for another teacher in my school who became interested when she observed the impact of the RTM on my teaching. Thus, as a high school mathematics teacher, I applied standards-based teaching in my classroom and mentored a peer through the RTM experience.

Serving as mentor, who possessed classroom-based RTM experiences, prepared me with a rich perspective and understanding of the challenges and risks that teachers take when implementing reformed practice. The experience taught me to be cautious in the mentoring process. I had to remind myself consistently that mentoring was not a process of cloning teachers to be like me, but rather a means for facilitating teachers in their own personal directions and assessments in standards-based practices. As our relationship grew in the experiences of mentoring, I tried to remain cognizant of the stance of a mentor—to nurture, to observe, and to simulate the teacher's ability to reflect by asking probing questions that fostered teachers' abilities to critically analyze their practice.

When I became a researcher in the university setting, I continued mentoring teachers in RTM practice. This is how Carmelita and I met. Carmelita volunteered to participate in a RTM summer professional development as one of 10 teachers whom I would mentor. During the summer professional development, I remember how excited Carmelita was when she engaged in paired problem-solving and case discussions. Her level of enthusiasm and engagement remained consistent whenever I visited Carmelita's classroom for plan-teach-debrief cycles.

However, it was not until Carmelita invited me to share in the development and teaching of the geometry unit from a project-based approach that we began to realize that our relationship was more than a mentor-teacher relationship. We wanted to examine what was happening in Carmelita's classroom and the impact of our collaboration in this process. We wanted to share our knowledge of our findings from this classroom and we wanted to make our findings public through publications and presentation. We believed that what we were doing needed to be shared with others who were searching for ways to improve mathematics achievement for urban learners.

As our partnership evolved into a research perspective, it became clear that my knowledge as a researcher created a hierarchical relationship in the mindset of Carmelita. Often Carmelita, would say, "You know how to conduct research and I have not learned any methods of research. So, you can do the research on my classroom." It was during these times that I felt uncomfortable, because I wanted us to interact as equal partners in research. My approach to changing Carmelita's mindset was to remind her of her knowledge base and rich experiences that were pertinent to the research. I reminded Carmelita continually that without her participation as a partner in the research, it would not be possible for me to access crucial data on her students, her classroom and her teaching. As Carmelita began to engage as a partner in research, her views of herself as researcher became apparent through her display of confidence in the research process. Finally, Carmelita claimed her status and value as a researcher. Our partnership was enriched by the diversity in our backgrounds and skills, as well as in our common understanding of, commitment to, and belief in reformed teaching for urban learners. Our partnership was based on trust in each other's abilities as researchers, respect for the work ethics of each other, the rigor and intensity that we brought to our work, and our beliefs in the mathematical abilities of urban learners.

SUMMARY

Our teacher-researcher collaboration proved to be a rich approach to making mathematics meaningful for urban learners. In this chapter, we examined the steps that we took in the evolution of our collaboration, and we uncovered the usefulness of our collaboration in the delivery of high-quality mathematics instruction. We shared our story of how we interacted as a teacher-researcher collaboration, how the project-based unit that was developed to provide an engaging and challenging approach to the curriculum, and how our work impacted urban learners in mathematics. We close with some of Carmelita's words:

> What we accomplished in this collaboration cannot be seen just as a unit in geometry or set of activities. When we take what we do to other teachers, I do not want them to think that we can just give them this unit or teach them how to build the icosahedra and all that we have as a teacher-researcher collaboration can or will happen in their classrooms. What we have is much more complex and sharing what we have requires much more than a workshop or demonstrating how to teach a mathematical activity. Teachers need to know how to reflect on their practice, how to analyze student work, and when and how to intervene. What we have as a teacher-researcher collaboration is complex and this message is what we must take to other teachers who truly are searching for ways to improve the mathematical performance of urban learners. Certainly, the statement that teaching is a complex endeavor and cannot be reduced to prescriptions or recipes rings true.

REFERENCES

Fetter, A. E., Schmalzried, C., Eckert, N., Schattschneider, D., & Klotz, E. (1991). *The platonic solids: Activity book*. Berkeley, CA: Key Curriculum Press.

Gardner, H. (1993). *Frames of mind: Theory of multiple intelligence* (10th anniversary ed.). New York: Basic Books.

Gardner, H. (1994). The multiple intelligence theory. In R. J. Sternberg (Ed.), *Encyclopedia of human intelligence* (Vol. 2, pp74–742). New York: MacMillan.

Hart, L., Schultz, K., & Najee-ullah, D. (1992). Implementing the professional standards for teaching mathematics: The role of reflection in teaching. *Arithmetic Teacher, 40*(1), 40–42.

Hart, L., & Najee-ullah, D. (1997). *The reflective teaching model* [CD-ROM]. Atlanta, GA: Galileo, Inc.

Hart, L. (2002). A four-year follow-up study of teachers' beliefs after participating in a teacher enhancement project. In G. C. Leder, E. Pehkonen, & G. Torner (Eds.), *Beliefs: A hidden variable in mathematics education* (pp. 161–176). Boston: Kluwer Academic Publishers.

Hart, L., Najee-ullah, D., & Schultz, K. (2004). The reflective teaching model: A professional development model. In R. Rubenstein (Ed.), *Effective mathematics teaching*. Reston, VA: National Council of Teachers of Mathematics.

Keys, C., & Golley, P. S. (1996). The power of a partner: Using collaborative reflection to support constructivist practice in middle grades science and mathematics. *Journal of Science Teacher Education, 7*(4), 229–246.

Kitchen, R. S., & DePree, J. (2005). Closing the gap through an explicit focus on learning and teaching. *National Council of Supervisors of Mathematics Journal*, 3–8.

Ladson-Billings, G. (1995a). But that's just good teaching! The case for culturally relevant pedagogy. *Theory Into Practice, 34*, 159–165.

Ladson-Billings, G. (1995b). Toward a theory of culturally relevant pedagogy. *American Educational Researcher Journal, 32*, 465–490.

National Council of Teachers of Mathematics (1989). *Curriculum and evaluation standards for school mathematics*. Reston, VA: Author.

National Council of Teachers of Mathematics (1991). *Professional standards for teaching mathematics*. Reston, VA: Author.

National Council of Teachers of Mathematics (1995). *Assessment standards for school mathematics*. Reston, VA: Author.

National Council of Teachers of Mathematics (2000). *Principles and standards for school mathematics*. Reston, VA: Author.

Pappas, T. (1989). *The joy of mathematics: Discovering mathematics all around you*. San Carlos, CA: Wide World Publishing/Tetra.

Simon, M. A. (1995). Reconstructing mathematics pedagogy from a constructivist perspective. *Journal for Research in Mathematics Education, 26*, 114–145.

Simon, M. A. (1997). Developing new models of mathematics teaching. In E. Fennema & B. S. Nelson (Eds.), *Mathematics Teachers in Transition* (pp. 55–88). Mahwah, NJ: Lawrence Erlbaum Associates.

Thomas, C. D., & Santiago, C. A. (2002). Building mathematically powerful students through connections. *Mathematics Teaching in the Middle School, 7*, 484–488.

Walker, P. C., & Chappell, M. F. (1997). Reshaping perspectives on teaching mathematics in diverse urban schools. In J. Trentacosta & M. J. Kenney (Eds.), *Multicultural and gender equity in the mathematics classroom: The gift of diversity* (pp.
201–208). Reston, VA: NCTM.

Ulichny, P., & Schoener, W. (2000). Teacher-researcher collaboration from two per-
spectives. In B. M. Brizuela, J. P. Stewart, R. G. Carrillo, & J. G. Berger (Eds.),
Acts of Inquiry in Qualitative Research (pp. 177–205). Cambridge, MA: Harvard
Educational Review.

CHAPTER 9

EXPLORING CONNECTIONS BETWEEN TEACHING, LEARNING AND ASSESSING MATHEMATICS FOR UNDERSTANDING

Tracey J. Smith
Charles Sturt University,
Wagga Wagga, New South Wales, Australia

Writing reflectively about your own practice is one way of making sense of experience so that it can be shared by others. In writing this chapter, I have attempted to capture some of the inherent complexities that I faced when I decided to change my approach to the teaching and learning of mathematics. My description takes the form of a narrative inquiry (Clandinin & Connelly, 2000) which seeks to piece together some of the insights I gained as a teacher researcher seeking to transform my classroom practice. Not only did I seek to teach mathematics for understanding, I also wanted to explore ways of assessing students' understanding of mathematics in more strategic ways.

Teachers Engaged in Research
Inquiry Into Mathematics Classrooms, Grades 6–8, pages 171–190

IMPETUS FOR THE STUDY

Having spent thirteen years as a classroom teacher in school settings in Australia and Canada (for one year) that ranged from Kindergarten to Grade 10, I then took on a new role of mathematics consultant for our state department of education. My role was to assist teachers in district schools to implement a new curriculum document that reflected reform visions for teaching and assessing mathematics similar to those in other countries (e.g., Black & Wiliam, 1998; National Council of Teachers of Mathematics [NCTM], 2000; Stenmark, 1991). At the same time, I had enrolled part-time in a university program to complete a post-graduate degree. The literature provided by the department to support my role as a consultant and the set readings for my courses began to extend my theoretical understanding about a variety of approaches for teaching and assessing mathematics. The university program also provided an opportunity for me to take a course in introductory research methods.

The new insights I gained as a mathematics consultant and a part-time student were enlightening and somewhat empowering, but my underlying thought was whether or not some of the new practices I had become more aware of would work in a classroom. During my first year back teaching after my two years as a mathematics consultant, I could hardly wait to put theory into practice to see if this vision of teaching, learning and assessing mathematics for understanding was going to be workable. One of the main tools for teaching and learning that I had become much more aware of was the use of open-ended tasks. In particular, the work of Sullivan and Lilburn (1997), Stenmark (1991) and Clarke (1996) had captured my interest because they characterized the use of open-ended tasks not only as a teaching and learning tool, but a way of integrating assessment with instruction. This was one area of my practice that I felt needed to be developed, so I decided to inquire into the extent to which open-ended tasks could be used simultaneously as a tool for teaching, learning *and* assessing for mathematical understanding.

For one of the graduate courses I was enrolled in, *Creative Classroom Mathematics*, I was able to determine my own focus of inquiry to complete the requirements of the course. As a consequence, I decided to research my own attempts at transforming my practice by collecting data that would help me understand the ups and downs of implementing open-ended tasks. My process of inquiry followed repeated action research cycles of plan, act, observe, reflect and then re-plan (Kemmis & McTaggart, 1988). These cycles varied in length from one lesson to one week over a six-month period. As I learned more about this process through my course at the university, it resonated strongly with what I believe most teachers already do quite naturally in their everyday practice. However, the role of a teacher researcher needs to shift towards more "systematic, intentional inquiry by teachers about their own school and classroom work" (Cochran-Smith & Lytle, 1993, p. 24). As a result of this shift, I became more attuned to notice particular events, but did not record these in a journal every day. Instead, I wrote notes in my plan book about significant events in the classroom, including which students' work samples had provided valuable insights, and reflected on these at the end of the week in my journal. I then photocopied these students' work so that I could analyze them further and use them as part of my assignments at the university. All of these data sources have provided the threads that are now woven together to form this narrative.

SCHOOL SETTING

The school where I was teaching when I collected the data was situated in a small rural town forty kilometers north of a large inland city in Australia. Wiradjuri Central School (pseudonyms are used throughout this narrative) had 300 students from Kindergarten (the first year of school) to Grade 12 (the final year of school). I was teaching a Grade 6 class that consisted of 28 students aged eleven or twelve years. The students were in their final year before they moved on to high school, although in this case because of the nature of central schools, the high school was situated on the same site as the primary school. The students had a diverse mix of socio-economic backgrounds and abilities ranging from some very committed learners to some very hesitant students who had given up on themselves as learners. For these students their self-esteem was extremely low and the expectation that they would understand what they were learning in mathematics was notably absent.

The reflections in this chapter are framed around my experiences teaching a unit on volume. The focus for the unit was to develop a conceptual understanding of the appropriate selection and use of the cubic meter and cubic centimeter as formal units of measure. This narrative seeks to describe the nature of my learning experiences as I set about transforming my practice. The unit on volume lasted about four weeks, but other areas of the mathematics curriculum were covered concurrently. To structure my narrative, I have used a framework related to teaching and learning for understanding outlined by Hiebert et al. (1997). The literature related to teaching and learning for understanding, including earlier work by Hiebert and Carpenter (1992), had played an influential role in my developing beliefs about the pedagogy of mathematics. I liked the idea that understanding could be thought of as a generative and emerging process instead of an all or none phenomena. Hiebert et al. (1997) suggested that to understand something, we need to "see how it is related or connected to other things we know" (p. 4) and be able to describe the relationships between those things. How could I transform my teaching to enhance such an understanding of mathematics?

Hiebert et al. (1997) described the critical features of classrooms that foster learning with understanding. According to these authors, the five dimensions of the classroom that are critical in supporting and nurturing understanding are: (a) the nature of the learning tasks, (b) the role of the teacher, (c) the social culture of the classroom, (d) the kind of mathematical tools that are available, and (e) the equity and accessibility of mathematics for every student. Each of these features is used as a framework for structuring and describing my experiences as a teacher researcher.

THE NATURE OF THE LEARNING TASKS

I see mathematics as one way of ordering and making sense of the world around us. As teachers, we should be opening up the process of learning mathematics to make room for multiple pathways for getting to a solution and to provide more time for students to ponder (reflect on) and share (communicate) those pathways. These beliefs are directly related to the nature of the tasks I planned during the unit on volume. It seemed important to me that learners should have opportunities to participate in tasks that are challenging in nature and framed around meaningful contexts while also being supported when necessary. As a consultant, I became very aware of the value of open-ended questions and tasks as a way of enacting these beliefs, catering for students' varying abilities and learning styles and making explicit students' thinking processes. But would they work in practice?

Using an Open-ended Approach

Open-ended tasks require students to do more than recall known facts and have the potential to stimulate higher levels of thinking (Clarke, 1996; Stenmark, 1991; Sullivan & Lilburn, 1997). I think that one of the most empowering tools for teachers seeking to transform their practice is the availability of quality resources to support them. The book written by Sullivan and Lilburn had become a valuable resource for my work as a consultant because it provided a diverse range of open-ended tasks to use with teachers and students from Kindergarten to Grade 10. By using this resource in my own classroom I was able to develop a sense of confidence in planning open-ended tasks and then, eventually, to create my own tasks by tweaking the tasks in the book to make them more appropriate for the context in which I was teaching.

In the classroom, I soon discovered that having an open-ended task at the beginning of a unit was an ideal way to establish students' background knowledge before I began a new topic. For example, to begin my unit on volume with Grade 6, we co-constructed a concept map that outlined background knowledge about the topic. To do this, I created a conversation by asking the question "What do we know about the term volume"? By reflecting on and recording our personal and collective resources, we were able to establish a shared vocabulary and a shared understanding of volume. The concept map that was created was then posted on the classroom wall to be added to at any time. What became clear as a result of developing the concept map was that the students possessed the language to describe units of measure for finding volume. The terms cubic centimeter and cubic meter were part of a common vocabulary, but as the conversation moved to predicting and estimating volume, I quickly determined that their mental models of these measures needed developing. To me, this was one of the first realizations of how an everyday instructional strategy such as a structured classroom conversation could become an opportunity for assessment.

THE ROLE OF THE TEACHER

Learning to Let Go

Incorporating a more open-ended approach to teaching and learning mathematics was clearly going to involve students letting go of their well-established views of mathematics and how it should be experienced. My own shift in thinking towards teaching for understanding meant that I was letting go of the role of teller and becoming more of a listener. I began to realize that listening more and telling less meant that I needed to plan more opportunities for students to verbalize, clarify and record their thinking so that they could reflect on and communicate their understandings more publicly. In this way, I could listen in and share aspects of their thinking, such as their ability to communicate and reason mathematically, that were often left hidden when they did more structured worksheets. What I tended to forget about was that listening more required students to think, talk and record ideas about mathematics more than they ever had before!

Learning *when* to let go and *what* to let go of became part of the struggle to understand what constituted an appropriate balance between explicit instruction and student-centered, inquiry-based learning. When should I tell students important facts and when should I let them find out for themselves? This sense of struggle to find a balance seemed to reflect a critical shift in thinking that, as a consultant, I saw many teachers grapple with.

What I noticed most about my practice was that I began to shift the moments of direct instruction so that they occurred towards *the end* of a lesson instead of at the beginning. As we collectively reflected on and shared conversations about existing knowledge, it became much clearer what direct instruction was necessary to move the students forward in their conceptual understanding. Once again the relationship between teaching, learning and assessing was clearly identified and the boundaries between them were becoming more blurred.

An illustrative example will help to explain learning to let go more clearly. As a result of doing the concept map mentioned earlier and assessing-in-the-moment by listening more attentively to student responses, I changed the plan for the next lesson to be an open-ended task that required students to form small groups and make a cubic meter using only newspaper, masking tape and a meter ruler. Such an open-ended task allowed for the development of a mental model of a cubic meter to build on new ideas about predicting and measuring the volume of larger spaces. The difficulties in visualizing a mental model of a cubic meter emerged during the development of our concept map when students' estimates of how many cubic meters it would take to fill our classroom were way off target. Figure 9.1 portrays the construction of a cubic meter by a group of students to help them visualize and get a feel for its size. This particular group also included a floor section to their cubic meter, which provided a mental model for a square meter as well.

Figure 9.1.   Constructing and visualizing a cubic meter.

Listening In on Conversations

Such an open-ended task became a fascinating thing to observe and provided an opportunity for me to listen in on students' conversations and learn about each student as they attempted to build the cubic meter. As I wandered between the groups, I noticed that all students were on task and completely engaged, an event that did not always occur in our classroom. This gave me the opportunity to eavesdrop on the conversations within the groups. In particular, I overheard Jon (a student with learning difficulties) taking charge of his group and saying "We've got to roll the newspaper out so that it's 100 centimeters long. This is only 70 centimeters long so we need another 30 centimeters to make it a meter." While I tried to hide my look of astonishment, it affirmed my belief that open-ended tasks undertaken in small collaborative groups was an ideal opportunity to find out about students' background knowledge because the nature of the tasks created a less intrusive context that required a more hands-on approach. Such an open-ended approach seemed to engage students like Jon more successfully.

In addition, as I listened in on their conversations I used anecdotal notes next to students' names on a class list to strategically record some of my new insights into students' thinking. I was able to write comments that related to students' ability to think and communicate mathematically such as "predicted how many of the cubic meters would fit in the classroom and justified her reasoning" and "noted that his estimate would have to change because the cubic meter is much bigger than he thought." These types of anecdotes became extremely useful resources for writing students' mid-year reports to send home to their parents because I was able to write more personal examples of their mathematical thinking. It became another instance when I could really see how assessment could become an integral part of instruction and more effective reporting.

Think-alouds

When we returned to the classroom after constructing the cubic meters, we revisited our predictions related to the volume of our classroom. Having constructed a visual image of a cubic meter, many students were able to see that their predictions needed modifying. As a result of overhearing Jon's conversation outside about finding volume, I asked him to come up and think aloud about his process for finding the volume of our classroom more accurately. Jon's think-aloud described a process of "imagining lines of the cubes we made lying next to each other across the room" and then he went on to say that "there would be layers of these lines all the way to the roof." During this think-aloud, I had to catch myself in the moment so I did not immediately introduce the formula for finding volume, which I have done many times in my past practice. This time, after Jon's think-aloud I asked the class if there was anything else we needed to consider and other students determined that there should be no gaps or overlaps during this process. This was an important step in developing students' understanding that volume can be thought of as structured arrays of horizontal and vertical layers rather than just collections of cubes. I then intervened with my own think-aloud by asking "I wonder if we would have to physically lay out the cubes or is there a more efficient way of finding volume?" This question was motivated by the fact that only one of our cubes was rigid enough to be brought into the classroom from outside.

In this example, think-alouds created the space for students to establish a shared understanding of the process of finding volume. By asking Jon and his peers to think aloud, I created an opportunity for conversations that would allow me to listen more to students' existing language and understandings rather than immediately presenting an abstract formula. This idea of listening more and telling less became the basis for one of the major shifts in my thinking about my role as a teacher. One of the constant questions I kept asking myself in the moment was "How can I hand the telling over to the students?" Even though this strategy was effective, it was clear that only a few students felt comfortable thinking aloud and presenting their thoughts to their peers. In the next section on the social culture of the classroom, I describe how I modeled the think-aloud strategy to enhance a culture of thinking in our classroom.

Integrating Assessment with Instruction

Something that became very obvious to me was that a change towards teaching mathematics for understanding that had an emphasis on communication and reflection would require a reciprocal change in the way I planned to assess students' progress. My exposure to the recent literature related to assessment in mathematics as a consultant and part-time student overwhelmingly supported the nexus between teaching, learning and assessing by suggesting that assessment should be:

- integral to, and naturally derived from good instructional practice;
- accessible, relevant and meaningful to all students so they can show what they know, understand and can do;
- on-going and cyclic in nature; and
- an indicator of students' thinking as well as content knowledge. (Black & Wiliam, 1998; Bright & Joyner, 1998; Clark, 1996; Kulm, 1994; Lajoie, 1995; Mathematical Sciences Education Board, 1993)

I wanted assessment to be formative (Black & Wiliam, 1998) so that it would provide opportunities for students to reflect on and communicate what they know, understand and can do on a day-to-day basis. In turn, this information could guide my instruction in the short term (in the moment or in the next lesson) and the longer term (future planning of the next unit to build on background knowledge). The end-of-topic tests that I had used in the past seemed very inadequate if I wanted assessment to be a formative practice. Increasingly, the creation of work samples was becoming my most valuable means of gathering evidence of students' progress and providing feedback that was focused on learning and thinking. I was also becoming more discerning about the differences between a worksheet and a work sample which are explored in the section on mathematical tools. Making productive connections between open-ended assessment tasks and everyday instruction seemed to be directly linked to the type of culture I needed to create in the classroom.

THE SOCIAL CULTURE OF THE CLASSROOM

Breaking the Mathematical Contract

Changing the nature of the tasks that I presented to students was not all smooth sailing. In seeking to let go of the role of teller, students had to develop a sense of authority and identity as problem solvers rather than just followers of procedures. In past years, the students who were highly competent at mathematics had been left alone when they finished quickly their mathematics worksheets and were often allowed to have free time until the others were finished. Providing tasks that had multiple answers, and asking students to reflect on and justify their solutions seemed like breaking all the rules and norms for doing mathematics. My vision of myself as a facilitator of learning, where students were responsible for reflecting on and communicating their thinking about mathematics, was being challenged directly by the responses of some of the students in our class. I was clearly breaking the "didactic contract" (Brousseau, 1986, cited in Clarke, 1996) that defines the obligations and expectations of students and their teacher.

It quickly became very clear that I needed to create a shared understanding of my role as a facilitator and supporter of learning, as well as the tenets of behavior and the standards of performance that would be expected both from me as the teacher and from the students. I had read about Clarke's (1996) suggestion that "assessment can serve as a powerful agent for communicating changing standards" (p. 346) that make explicit the tenets of excellent mathematical performance. Using an open-ended approach to teaching and learning required students to become more active and communicative in their role as learners and to become more critical thinkers (reflective) about their learning.

As I reflected on my planned actions and the reactions of students to open-ended tasks, I realized that a commitment to an open-ended pedagogical approach would require a great deal of resilience to keep trying something I sensed was valuable but which was meeting definite resistance! Researching my own practice required me to continually reflect on my beliefs because the complex and uncertain nature of open-ended tasks seemed to be challenging everybody, including me. Was an open-ended approach asking too much of the students or was it worthwhile persevering? My strong belief was that an open-ended approach would build students' self-esteem and confidence in mathematics because they could gain access to open tasks at whatever level they felt comfortable. I decided to continue the next cycle of plan, act, observe, and reflect – resilience in action! Upon reflection, I could see more clearly that I needed to take the students with me along on the journey towards teaching and learning with understanding. I wrote in my journal at the time that "sometimes, I can be a slow unlearner!" Some of the strategies I used to take the students along with me so that we could establish a shared understanding about a new contract were to change my questioning techniques, model the think-aloud strategy myself, and give students a second chance to improve the way they had recorded their thinking.

Start Small, Think Big

I remembered that when you try something new you should "start small, but think big." Consequently, I began to change subtly the nature of my classroom questions by asking questions that started with "how" and "why." This strategy began to change the nature of the conversations that took place in our classroom. Another method I used was to insert the phrase "You know that because…" into my vocabulary. By showing a genuine interest in how students came to a solution and why they felt their answer made sense, the students' ideas became the currency of the classroom (Hiebert et al., 1997) and were respected for their potential to increase the collective understanding of the class (including me as their teacher).

Modeling the Think-aloud Strategy

The previous anecdote about Jon's think-aloud after we built our cubic meters outside was one example of how such a strategy helped to develop the thinking culture of the classroom. However, it was clearly a strategy that only a few students were comfortable with. I developed this strategy further towards the end of the volume unit when I set an integrated literacy task that required students to write a procedure for finding the volume of a rectangular prism. This time, I modeled the think-aloud strategy by stating that "If I was writing a procedure, I would have to think about the goal I was aiming for, the materials I would use, and the steps I would take to reach my goal. I would also want to use all the mathematical terms I could to help describe the process in detail." By modeling these thoughts about what should be included in a procedure, and writing these thoughts on the board, I was also making explicit my expectations of what students' responses might look like. The students also knew that this was one of their pieces of work that I would be collecting to determine how well they understood the concept of volume. Once again, assessment became an integral part of instruction. Placing thinking up front in our classroom conversations by using language such as "how," "why," "because" and "I think" helped to model a thinking language which in turn gave students an extended vocabulary to verbalize, clarify and record their thinking more easily. It also seemed to develop a more accepting culture within the classroom that valued each student's contribution.

A Second Chance at Learning

The writing of the procedures for finding volume led to another strategy that seemed to positively affect the way students viewed mistakes and helped to shatter the image that to be good at mathematics you need instant answers. After the Grade 6 students had written their procedures, I was disappointed to read them and find that six students recorded the process and told me to *add* the length and breadth and height of the container. While their work samples did demonstrate that they understood the need for three measurements to find volume, it was obvious that some revisiting of the process was necessary. This time, assessment led to the notion of *a second chance at learning*.

   Out of frustration, I decided to give the procedures back to all the students the next day without any feedback. I asked them to re-read their work and make any changes they thought were necessary to enhance their representation of what they knew about volume. This proved to be successful as four out of the six students I mentioned earlier noticed their mistake and corrected their work. In addition, other students who had already written a correct procedure added more details to their responses. While the rest of the class worked on a new task, I revisited the process with the two students who did not change their work. By giving students' work back to them with the expectation that they would think critically about their answers, they became the assessors of their own learning and mistakes became sites for a second chance at learning. I had read about Hiebert et al.'s (1997) idea that "mistakes must be seen by the students and the teacher as places that afford opportunities to examine errors in reasoning, and thereby raise everyone's level of analysis" (p. 9). It seemed that giving the students the opportunity for a second chance at learning did have the potential to shape their conceptions about mathematics. This practice also challenged the notion that mathematics solutions have to be determined quickly with no chance to reflect on those solutions in order to learn from them.

MATHEMATICAL TOOLS AS LEARNING SUPPORTS

According to Hiebert et al. (1997), mathematical tools should be used to support learning and solve problems. One of the goals for transforming my practice was to use a variety of written and oral ways of communicating and reflecting about mathematics as pedagogical tools. I used concept maps, work samples, written procedures, and oral reasoning (such as think-alouds) as well as physical materials such as base-ten blocks, construction materials and large plastic cubes to broaden my approach. While using open-ended tasks enabled students to think and find solutions in diverse ways, an important shift in my thinking occurred in relation to the term "problem solving." Even though open-ended tasks obviously required the solving of a problem, I stopped using the term "problem solving" with my Grade 6 when I noticed that they seemed to have an aversion to problems. On the other hand, challenges, investigations or tasks seemed to gain and maintain their interest more readily. A small, but important shift in my thinking seemed to make a big difference to students' attitudes and willingness to explore multiple solutions!

Worksheet or Work Sample?

In the past, I had been very accustomed to using worksheets as a teaching tool. For me, the term worksheets refer to sheets of work I may have photocopied from teacher resource books or textbooks that would have a large amount of text and/or images already on the paper. Sometimes, I would cut and paste from a number of different books to develop a teacher-made worksheet. While the students were required to complete the sheets, the information the completed sheets provided to me was often limited in terms of establishing students' background knowledge and thinking processes involved in completing the sheets. When I tried to reason about what students might be learning from the worksheets, I realized that the potential for worksheets to provide evidence of thinking processes such as reflecting, reasoning and communicating mathematically was quite limited.

In contrast, work samples allowed for a more open response to a task because there is very minimal, if any, text and/or images on the page to begin with. Setting more open-ended tasks and asking students to record their solutions, meant that the information I gained from a work sample was constructed solely by the student, which then provided more detailed insights into students' perceptions and thinking about concepts such as volume. I also learned over time that I could find out much more about students' thinking if I tweaked a task further and asked them to justify their solution (using the how, why and because questions). In this sense, the nature of open-ended tasks required the creation of work samples that provided a rich source of students' representations of their understanding. They also represented another form of letting go of the authority and ownership of what constitutes understanding. The creation of a work sample provided the opportunity for students to clarify and record their thinking, which led to assessment opportunities that provided "a window to students' thinking and a compass for instruction" (Cooney, Badger & Wilson, 1993, p. 239). This quote had become one of my favorite phrases while I was working as a consultant, and I was now playing it out in practice. I consider the two figures in the next section, constructed by Cara and Oscar, to be illustrative examples of work samples rather than worksheets.

A further example of using a variety of mathematical tools is drawn from the unit on volume where we explored the idea that as a unit of measure, volume is "related" to other units of measure. I introduced the idea that mass and capacity are "cousins" of volume and by using clear containers, metric scales and water to explore these relationships, the students were able to reason that one liter of water has a mass of one kilogram and a volume of one thousand cubic centimeters. Therefore, they were able to infer that one milliliter of water (1 ml) would have a mass of one gram (1 g) and its volume would be 1 cubic centimeter ($1cm^3$). This aspect of the unit called for a more structured and guided exploration rather than open discovery. Part of getting the balance right meant that I wanted to scaffold the students throughout this process so I could support them in making the connections among the different units of measure. This was necessary because students would need to understand the relationship between volume, capacity and mass in the metric system to undertake the final task in our unit on volume.

The final task provides an illustrative case of how the use of open-ended tasks as a pedagogical tool can create simultaneous opportunities for teacher and student learning and assessment. The final task was worded as follows: *Design a container that will hold three liters of milk and still fit in a refrigerator. You will need to justify your choice of design and present it clearly.* Assessment criteria for the task were shared with the students so they knew I was looking for evidence of the relationship between volume and capacity as well as documentation that was presented in an organized way to show thinking and reasoning. After collecting the students' work samples, I was very disillusioned to see the most wonderful designs that did not equal 3000 cubic cm in volume or 3000 ml in capacity! For many students, there appeared to be little transfer of knowledge to find the correct measurements. However, a number of students were close to the right size although they did not check or verify their solutions. The work samples I collected from students showed varying responses to the task. Figure 9.2 shows Cara's creative response that was mathematically incorrect – "the volume of the milk container is 8625 cm^3," and showed no verification for her design except that she "made it like a really small jerry can" so that it would fit in the fridge.

On the other hand, Oscar, managed to design and make a correct-sized container and justify it by showing his work. Figure 9.3 shows Oscar's less creative response that was mathematically correct and included reasoning for his response. As a result of most of the students' less-than-accurate

Figure 9.2.   Cara's response to the three-liter task.

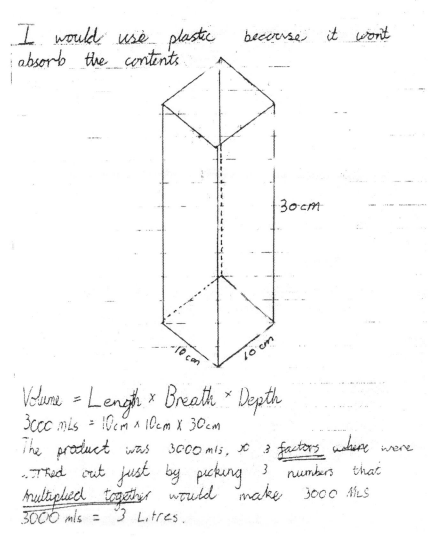

I would use plastic because it wont absorb the contents

Volume = Length × Breath × Depth

3000 mLs = 10cm × 10cm × 30cm

The product was 3000 mls, ∞ 3 <u>factors</u> <u>where</u> were worked out just by picking 3 numbers that <u>multiplied together</u> would make 3000 MLS

3000 mls = 3 Litres.

Figure 9.3.   Oscar's response to the three-liter task.

responses, I decided to use Oscar's work sample as an exemplary response to the task (with his permission). As a class we looked at his work sample on a large screen and discussed what aspects of his work sample made it clear that he understood the relationship between volume and capacity and that he had thought about the strategies involved in the process. The technique of using students' work samples to model best practice emerged as an extremely valuable tool for developing more detailed responses to open-ended tasks. It was another way of helping students come to terms with what was expected in our new contract for learning through open-ended tasks. In addition, during the week we did that task, the local Dairy Farmers company released a new three liter container of milk. This release made the news and a few of the students were very excited to see that they had been given a task to solve that actually happened in real life!

EQUITY AND ACCESSIBILITY

There is a strong nexus between the four other dimensions of classrooms and the fifth one related to equity and access. I believe that the open-ended, inquiry-based nature of the tasks I have outlined in my narrative created opportunities for students to take ownership of their learning and show their thinking in explicit ways. Whatever solutions they came up with were the result of their own construction of meaning and they presented us all with opportunities to learn together. My role as a facilitator of learning needed to be negotiated with students so that a shared understanding of expectations and valued practices could be made explicit. By providing a classroom culture that is supportive, conversational and respectful, each student hopefully felt valued for their contributions and more importantly, felt like they had a contribution to make. In addition, the fact that there were multiple pathways to access and think about open-ended tasks allowed every student in my class to participate in the same task to varying degrees. Open-ended tasks became a logical and effective way to cater for the diverse range of learners in our class.

REFLECTING ON MY LEARNING

The process of researching my own practice required me to become more attuned to the beliefs that were underpinning my practice. It also required me to not only see what happens when theory meets practice, but to become more intentional in taking account of my experiences by collecting data that was directly related to my practice. In this way, becoming a teacher researcher and writing a narrative about my experiences allowed me to theorize my own practice so that my personal theories of teaching, learning and assessing could be made more explicit. In terms of the inquiry focus, I have increased my understanding of how open-ended tasks can be used simultaneously as a tool for teaching, learning and assessing mathematics for understanding. Using open-ended tasks necessitated a transformation in the way I viewed my role as a teacher and the nature of the classroom culture that was established.

In summary, the personal theories I have developed as a result of researching my own practice suggest that the goal of connecting the processes of teaching, learning and assessing mathematics for understanding was enhanced by the use of open-ended tasks because they provided opportunities for:

- *Verbalizing* – asking "how" and "why" questions, the use of think-alouds by students and teachers, creating conversations to develop a shared understanding of a topic;
- *Clarifying* – establishing norms for what is valued in the classroom, negotiating clear assessment criteria that flag expectations, creating class or individual concept maps; and
- *Recording* – representing understanding through written procedures, concept maps, other work samples that include drawings, diagrams, tables or charts and justification of solutions to open-ended tasks.

These processes describe opportunities for extending the meaning-making process through the sharing of conversations that led to opportunities for assessing understanding in a more natural and on-going way. The processes of verbalizing, clarifying and recording provided opportunities for students to think, reflect on and communicate what they know, understand and can do in relation to mathematical concepts. In turn, this provided equitable and accessible opportunities for assessing students' understanding. Using these identified processes in subsequent research contexts, I have been able to develop a framework for promoting mathematical thinking and understanding in mathematics classrooms (Smith, 2000).

THE NATURE OF ASSESSMENT PRACTICES –
AN ADDITIONAL FEATURE OF CLASSROOMS

In creating my narrative, I have tried to illuminate how the process of researching my own practice led to a growing awareness of how the use of open-ended tasks supported the interdependent nature of teaching, learning and assessing mathematics for understanding. While the core features of classrooms outlined by Hiebert et al. (1997) provided a valuable framework to guide my narrative, these features did not fully account for the fact that a major corollary of teaching for understanding and learning for understanding is *assessing for understanding*. As Clarke, Clarke and Lovitt (1990) suggested some time ago, "it is through our assessment that we communicate most clearly to students which activities and learning outcomes we value" (p. 118). Above all, researching my own practice confirmed that if we value communicating ideas and thinking about why an answer makes sense, and if we truly believe that conceptual understanding is an important goal for students, then both our teaching *and* our assessment practices must simultaneously reflect those beliefs if we are to transform our future visions of practice. Perhaps this indicates that the nature of assessment practices should become an additional dimension of classrooms that support teaching and learning for understanding. In this way, assessment can be placed explicitly in the foreground of pedagogical practice alongside teaching and learning for understanding.

REFERENCES

Black, P., & Wiliam, D. (1998). Assessment and classroom learning. *Assessment in Education, 5*(1), 7-73.

Bright, G. W., & Joyner, J. M. (Eds.). (1998). *Classroom assessment in mathematics: Views from a National Science Foundation working conference.* Lanham, MD: University Press of America.

Cochran-Smith, M., & Lytle, S. L. (1993). *Inside/outside: Teacher research and knowledge.* New York, NY: Teachers College Press.

Cooney, T. J., Badger, E., & Wilson, M. R. (1993). Assessment, understanding mathematics, and distinguishing visions from mirages. In N. Webb & A. F. Coxford (Eds.), *Assessment in the mathematics classroom, 1993 Yearbook of the National Council of Teachers of Mathematics* (pp. 239-247). Reston, VA: National Council of Teachers of Mathematics.

Clandinin, D. J., & Connelly, F. M. (2000). *Narrative inquiry: Experience and story in qualitative research.* San Francisco, CA: Jossey-Bass.

Clarke, D. (1996). Assessment. In A. J. Bishop (Ed.), *International handbook of mathematics education* (pp. 327-370). Dordrecht, The Netherlands: Kluwer.Clarke, D. J., Clarke, D. M, & Lovitt, C. J. (1990). Changes in mathematics teaching call for assessment alternatives. In T. J. Cooney & C. R. Hirsch (Eds.), *Teaching and learning mathematics in the 1990s,* 1990 Yearbook of the National Council of Teachers of Mathematics (pp. 118-129). Reston, VA: National Council of Teachers of Mathematics.

Hiebert, J., & Carpenter, T. P. (1992). Learning and teaching with understanding. In D. Grouws (Ed.), *Handbook of research on mathematics teaching and learning* (pp. 65-97). New York: Macmillian.

Hiebert, J., Carpenter, T., Fennema, E., Fuson, K., Wearne, D., & Murray, H. (1997). *Making sense: Teaching and learning mathematics with understanding.* Portsmouth, NH: Heinemann.

Kemmis, S., & McTaggart, R. (Eds.). (1988). *The action research planner* (3rd ed.). Melbourne, Australia: Deakin University Press.

Kulm, G. (1994). *Mathematics assessment: What works in the classroom.* San Francisco, CA: Jossey Bass.

Lajoie, S. P. (1995). A framework for authentic assessment in mathematics. In T. Romberg (Ed.), *Reform in school mathematics and authentic assessment* (pp. 19-37). Albany, NY: State University of New York Press.

Mathematical Sciences Education Board & National Research Council. (1993). *Measuring what counts*. Washington, D.C: National Academy Press.

National Council of Teachers of Mathematics. (2000). *Principles and standards for school mathematics*. Reston, VA: Author.

Smith, T. J. (2000). Bridging the research-practice gap: Developing a pedagogical framework that promotes mathematical thinking and understanding. *Mathematics Teacher Education and Development, 2*, 4-16.

Stenmark, J. K. (1991). *Mathematics assessment: Myths, models, good questions, and practical suggestions*. Reston, VA: National Council of Teachers of Mathematics.

Sullivan, P., & Lilburn, P. (1997). *Open-ended maths activities - Using 'good' questions to enhance learning*. Melbourne, Australia: Oxford University Press.

CHAPTER 10

THE 1998 WINTER OLYMPICS

Documenting and Analyzing Student Learning in an Inquiry Approach

Cynthia H. Callard
Twelve Corners Middle School,
University of Roshester, Rochester, NY

MY JOURNEY AS A MATHEMATICS TEACHER

I began teaching middle school mathematics in the fall of 1988 in an affluent suburb of a western New York city. I was just out of graduate school
and excited to share my knowledge and love of mathematics with my students. After the initial tenuousness of beginning this new job, I quickly
established a comfortable routine for my classes: (a) as students entered
the room they completed a brief warm-up exercise reviewing concepts/
procedures that had been addressed the day before, (b) they checked their
answers to the previous nights' homework with an answer key on the overhead, and (c) I then presented a 20-minute lesson where I recorded
"important" mathematical terms, ideas and examples on the chalkboard
and the students copied this information into their notebooks. My goal

Teachers Engaged in Research
Inquiry Into Mathematics Classrooms, Grades 6–8, pages 191–218
Copyright © 2006 by Information Age Publishing
All rights of reproduction in any form reserved.

was to present the material to my students in a clear, concise manner and to avoid confusion. Occasionally, a student or two might ask a question about a particular step or ask me to clarify something that I said, but in general that was the extent of the students' engagement in the lesson. I assigned practice problems for homework. At the end of a chapter, students would take a test that contained problems similar to those we had done in class or on homework. If students did not perform well on the assessment, I attributed it to their own lack of preparation.

I was comfortable with this structure and organization of my class and this presentation of mathematics. It was how I was taught and it represented my views of mathematics—it was linear, step-by-step and we were all working in the same way towards the same answer. I became part of a long history of mathematics teachers presenting mathematics as isolated bits of facts and procedures passed along to the few passive students who were willing/able to take it in. I also received positive feedback and support from district administration, parents and from colleagues in the department as I was covering the material that I was responsible for and it seemed that most of my students were successful.

However, over the next few years—as I reflected on my practices and my students' learning of mathematics—I began to realize some major flaws with the belief system on which this model of teaching was based. I began to question things such as: "What big mathematical ideas are my students really learning?"; "What kinds of skills and strategies are my students able to apply outside of my classroom, when inside my classroom everything is neat and simplified?"; "What do I know about my students' mathematical thinking and understandings?"

One impetus for me asking these questions was the publication of the National Council of Teachers of Mathematics' (NCTM) (1989) *Curriculum and Evaluation Standards for School Mathematics*. Even though the curricular and pedagogical practices the *Standards* described were not the practices taking place in my classroom, the ideas within this document made sense to me. I agreed that students should be able to reason mathematically, problem solve, communicate mathematically, have confidence in their ability to do mathematics, and make mathematical connections. I began attending NCTM national and regional meetings. I gathered activities I felt would help my students to meet the goals presented in the *Standards* and I began inserting them within my otherwise traditional mathematics course. However, although the *Standards* provided me with a set of goals and a vision for school mathematics that I believed in, I found it very difficult to figure out how to put the goals and rhetoric into practice in my classroom.

A large part of my growth and development as a mathematics teacher began in 1993 when I became involved in a National Science Foundation-funded project through the University of Rochester (Woodward, Borasi &

Packman, 1991). Through involvement in this project, my beliefs about the teaching and learning of mathematics, and as a result, my classroom practices, began to change significantly. This project promoted and supported the development of classroom instruction based on an *inquiry approach* to mathematics instruction as articulated by Borasi (1992, 1996). An inquiry approach is consistent with the goals and practices proposed in the *Curriculum and Evaluation Standards* (NCTM, 1989) and the subsequent publication available at the time, the *Professional Standards for Teaching Mathematics* (NCTM, 1991). Adopting an inquiry approach calls for a fundamental shift in assumptions about mathematics, teaching and learning consistent with the *Standards* documents, that is, a shift from a transmission theory of learning which informed most school mathematics instruction (as illustrated in the description of my classroom 15 years ago) to a constructivist theory of learning (Confrey, 1990; Davis, Maher, & Noddings, 1990; Simon, 1995). A constructivist theory of learning calls for students to develop mathematical understandings through active engagement within rich mathematical experiences where they explore big mathematical ideas. A constructivist theory of learning underlies the NCTM's vision for school mathematics and Borasi's inquiry approach to mathematics instruction.

INQUIRY-BASED INSTRUCTIONAL UNITS

Supported by a team of mathematics educators, researchers and teachers in this NSF-funded project, over the next several years a colleague and I developed and implemented three inquiry-based instructional units for eighth grade mathematics (hereafter referred to as "inquiry units"). Each unit was four to eight weeks in length and focused on a different broad mathematical topic. These units were specifically designed to address the NCTM *Standards* (1989, 1991, 1995) and the New York State *Learning Standards for Mathematics, Science, and Technology* (New York State Education Department, 1996).

The first inquiry unit, "Developing Area Formulas," addressed mathematical content in the areas of algebra, geometry and measurement. The focus was on the concept of area, the use of area formulas as a way (but not the only way) to determine area, and the process of developing area formulas. Traditionally, the focus of an eighth grade unit on area would be on the memorization and procedural application of formulas to calculate area. In contrast, in this unit students were expected to develop their own strategies and formulas to determine the area of standard and non-standard figures. Approaching area from this perspective allowed students to *act as mathematicians* and encouraged them to view mathematics as the product of human activity.

The second unit, "Learning Geometry Through Tessellations," was designed using the theme of tessellations as a context for students to investigate geometric shapes and their properties. In addition, students reflected on the nature and role of definitions in mathematics and engaged in an independent research project, developing and testing their own conjectures about what figures tessellate. This inquiry unit pushed student thinking well beyond the conventional curricular expectations of identifying and defining geometric terms and properties.

The third inquiry unit was designed as a thematic unit around a current event, the 1994 and later the 1998 Winter Olympics. The unit focused on the mathematical content areas of measurement and data representation and analysis, which had not traditionally received much attention at eighth grade. Throughout this unit, students developed and researched mathematically interesting questions about the Olympics and used mathematical tools to collect and analyze data to answer their questions. This chapter will focus on the description of student learning in this particular unit.

Consistent with the principles of an inquiry approach, all of these units were designed to involve students in exploring mathematics, discussing and writing about mathematics, working in groups and on projects, and presenting results to the class. Specific activities within these units were developed to help students realize the value of mathematics and to make connections between mathematics and other disciplines. The units were also designed to capitalize on student strengths so as to help students become more confident in their ability to do mathematics.

Based on our informal reflection on students' engagement in these units and on their written work, my colleague and I felt that as a result of participating in these three inquiry units our students were successful in terms of learning mathematical content, demonstrating a use of mathematical processes and displaying a more positive attitude toward mathematics. These results, however, had not been formally documented. Therefore, we did not have the necessary systematic research to share our findings with others in the field. The need to engage in systematic research led me to the dissertation study on which this chapter is based (Callard, 2000). My goal was to document and analyze what students can learn in a mathematics classroom informed by an inquiry approach.

INSTRUCTIONAL CONTEXT AND RESEARCH DESIGN

The experiences and research results described in this chapter took place during the 1997–1998 school year in an eighth grade mathematics classroom in which I was the mathematics teacher and the researcher. The school district is located in an affluent suburb of a western New York city and has a high academic orientation. The school district and community place a great deal of emphasis on—and pride themselves on—students' academic performance, particularly in the areas of mathematics and science.

Students in this sixth through eighth grade middle school were tracked for mathematics and science—that is, classes were homogeneously grouped according to ability as determined by teacher recommendations and standardized test results. There were three such mathematics tracks at the eighth grade level at the time of this study: (a) an "accelerated" mathematics class where eighth grade students took the mathematics course typically offered in ninth grade (approximately 35% of eighth graders), (b) "regular" Math 8, and (c) "blended" or "modified" mathematics for special education students mainstreamed in regular eighth grade mathematics with a special education teacher in the classroom (approximately 4% of eighth graders). Other labeled special education students were mainstreamed in regular eighth grade mathematics classes and received support from a learning disabilities specialist at a different time in the day, often in many different subject areas. Other non-labeled students who experienced difficulties in mathematics could receive support from a school tutor outside the classroom. Students whose primary language was not English could also receive support from a specialist at a different point in the school day in areas in which they might be having difficulty. Thus the regular Math 8 classes themselves included a wide range of students and ability levels.

I selected one of the three regular eighth grade math classes that I taught during the 1997–1998 school year to participate in this year-long study. This class was comparable to the other two regular Math 8 classes that I taught that year in terms of male/female ratios, number of students with special needs and included students with a wide range of abilities. More specifically, for the implementation of this particular unit, the class consisted of 10 female students and 10 male students. Of these 20 students, one student was labeled as a special education student and received support for all subject areas from a learning disabilities specialist once a day outside of mathematics class, and two other students (one female, one male) were not labeled but received support from a school tutor every other day for all subject areas.

As noted previously, the primary purpose of this study was to document and analyze, in a systematic way, students' learning in a mathematics classroom informed by an inquiry approach. In order to document students' learning within inquiry units formally, however, I had to come to a deep understanding of issues of assessment related to how to best measure students' learning within an inquiry approach. According to Webb (1992), "fundamental to mathematical assessment is an explicit definition of the content to be assessed" (p. 664). This specification of content is complex and is "derived from the purpose for assessment, a conception of mathematics, and a theory for the learning of mathematics" (p. 665). This highlighted the necessity of articulating up front what I felt was important to assess. However, in order to begin to consider *what* I wanted to assess and *how* I could go about this assessment, I needed to consider first my goals for students' learning.

Using the NCTM and NYS *Standards* documents and the curriculum guide established by the school's mathematics department, I carefully and clearly articulated goals for my eighth grade students' learning within the inquiry units previously described. I separated these goals into three categories: (a) goals for mathematical content, (b) goals for mathematical processes, and, (c) goals for mathematical disposition. In this chapter, I will focus on the documentation and analysis of students' learning of mathematical content goals for one particular unit, "Investigating the 1998 Winter Olympics."

In this unit the goals for students' mathematical content learning included goals regarding *specific mathematical knowledge* in the areas of measurement and data representation and analysis. I separated these goals into two categories: *factual content goals* and *"big idea" content goals* in order to distinguish those that focused on more straightforward applications of facts and procedures from those that focused on the understanding and use of broader mathematical ideas and concepts. For example, within the data representation and analysis content strand I considered "Students construct, read, and interpret tables, charts, and graphs" (NCTM, 2000) as a *factual content goal*, whereas I considered "Students formulate meaningful questions and collect, organize, display and analyze appropriate data to answer these questions" (NCTM, 2000) as a *"big idea" content goal*.

Throughout the unit, I audio taped each day's lesson and collected all written student work. I analyzed this data in order to draw conclusions about each student's learning of each content goal. Before discussing the details of the documentation and analysis of student learning in the unit, I provide here a brief overview of the inquiry unit under examination, "Investigating the 1998 Winter Olympics."

"INVESTIGATING THE 1998 WINTER OLYMPICS" UNIT GOALS AND DESCRIPTION

This unit was designed as a thematic unit around a current event, the 1998 Winter Olympics. In this unit, I capitalized on the students' interest in the Olympics to develop an appreciation of the use of mathematics in real-life situations, to introduce mathematical tools, and to provide students an opportunity to inquire on a question of their own. This unit focused on the mathematical content areas of measurement and data representation and analysis, and lasted approximately five weeks. The mathematical content goals for this unit can be found in Figure 10.1.

The description of this unit is divided into five main components (although these components did not take place in a linear fashion, but rather overlapped with each other): (a) gathering students' initial ideas and interests, (b) modeling the process of testing a conjecture using data, (c) setting the stage for independent inquiries, (d) other teacher-initiated investigations, and (e) culminating activities—independent research project, poster project and sharing of results, unit portfolio and reflection.

Gathering Students' Initial Ideas and Interests

In order to spark students' interest in this new unit, the unit began with students viewing a video clip that highlighted various Winter Olympics sports. Skiers, skaters, and lugers were shown performing flips, turns and slides in a brief preview of some of the Winter Olympic sports. The students

---

Measurement

1. Students develop a broad understanding/appreciation of the use of formulas to determine measurements (*"big idea"*)
2. Students appreciate what is involved in measuring and gain an understanding of <u>what</u> and <u>how</u> to measure for a specific purpose (*"big idea"*)
3. Students estimate and perform measurements in a variety of situations to describe and compare objects and data (*factual*)

Data Representation and Analysis

4. Students formulate meaningful questions, collect, organize, display and analyze appropriate data to answer these questions (*"big idea"*)
5. Students construct, read, and interpret tables, charts, and graphs (*factual*)
6. Students calculate and compare measures of central tendency (*factual*)

Figure 10.1.   Mathematical content goals for
             "Investigating the 1998 Winter Olympics."

were then given a questionnaire that included five "rich questions" designed to engage them in thinking and talking about the Winter Olympics (see Figure 10.2).

Students responded to these questions individually first, then with a partner, and finally in a whole-class discussion. These questions generated a great deal of excitement and interest in the Winter Olympics, particularly the question "In which event do the athletes go the fastest?"

To gather further information that would help in my planning, I asked the students to complete a survey for homework on the first day of the unit. This survey asked students to record Winter Olympic sports in which were they interested, their preferences for other students in the class to work with, and questions that they had about the Winter Olympics. Based on the students' interests and preferences of people with whom they wished to work, I assigned them to groups of two or three and assigned them a sport that they were expected to become "experts" in throughout the unit.

Modeling the Process of Testing a Conjecture using Data: Investigating the Question "In which Olympic Sport do Athletes go the Fastest?"

Based on students' unexpected interest in the question, "In which Olympic sport do athletes go the fastest?", I modified my original plans in order for the students to explore this question further. This exploration began spontaneously when two students, unprompted by me, gathered some contradictory data about the speeds of a few Winter Olympic sports. This information was used as a catalyst for further exploration of speed, what it means to measure speed and what is meant by average speed.

---

1. How can we compare the performance of the different countries participating in the Olympics?
2. In which event do the athletes go the fastest? Explain your answer.
3. Do you think the performance of the athletes is improving in each Olympics? Explain and give examples from specific events.
4. Do you think age is a factor in winning Olympic events? Explain.
5. How is math used in the Olympics? Provide specific examples in your explanation.

Figure 10.2.   Winter Olympics questionnaire.

In order to clarify some of the issues that arose during this discussion, I orchestrated a race between two students. We collected data consisting of the times it took each student to run different intervals—up a set of stairs, around a corner, and down a straight hallway—of a racetrack. We then used this data to calculate average speeds over different intervals as well as the average speed for the entire race. Using this concrete, and more simplistic, model helped us to examine some of the complex issues related to measuring and reporting speeds.

We then returned to our original question, which we realized we needed to modify somewhat. We realized that we would need to consider only average speeds since we did not have the data necessary to calculate intermediate speeds for the various events. We also realized that we were going to need to make some more decisions about what data to use, whether to use a specific medal winner or some sort of average, which year to collect our data from, and in what units we should record our data. After making these decisions, the students performed calculations to determine average speeds within each of the Winter Olympic sports in meters/second. Since the initial data that students had gathered were in miles/hour, we then worked on developing strategies to convert meters/second into miles/hour.

In order to bring some closure to the many discussions and experiences we had working with the issue of speed, students were given a homework assignment asking them their thoughts, ideas, and lingering questions about speed. Also, as a culminating activity for this investigation, as well as an individual assessment of some of the mathematical concepts we had addressed, the students were presented with a "compulsory exercise" (a "teacher-initiated" investigation) in which they had to compute the average speed of the top five athletes in their chosen event for the 1998 Winter Olympics, both in meters/second and miles/hour. In order to review some basic statistics, this task also asked students to calculate the mean, median and range of these values and to provide other information about speeds in their sport from newspaper articles, book resources, or class handouts and compare these with the speeds they had calculated. I modeled the use of a spreadsheet as a tool the students could use to calculate these speeds and perform the conversions.

This investigation was also used as a context for students to use various types of graphs (e.g., bar graphs, line graphs) and tables to report and analyze data. I also introduced the students to scatter plots as a way to investigate relationships between two variables in order to consider some of the factors that affect speed.

In order to reflect on the process that we had engaged in as a class to answer our initial question "In which event do the athletes go the fastest?" and as a model for the students' independent research project that they would engage in later in the unit, the students generated, as a class, a list of "Key Steps in Testing a Conjecture Using Data" summarizing our process (see Figure 10.3). In our discussion, we noted the cyclical and complex nature of the process, and limitations with data that they could encounter in their independent project as well.

Setting the Stage for Independent Inquiries

Following this rich investigation, students were ready to begin investigating a research question of their own. In order to introduce this activity, I shared a list of possible questions for students to investigate based on questions that students had raised throughout the unit so far (see Figure 10.4). Each group of students then decided on a research question that they were interested in investigating, or created a new question, based on the sport in which they were becoming an "expert".

I encouraged students to gather information and data from newspaper and magazine articles as well as to search the internet for information related to their question. A collection of books on the Winter Olympics

1. Rich question: "Who's the fastest?"
2. Shared ideas, discussed, debated—made our first conjectures.
3. Gathered information that caused some disagreement—more "arguing" because data was unclear.
4. Raise more questions, investigate further, did an experiment in order to define what we meant by speed!
5. Gathered more data on distance and time—made decisions about data
6. Specific question: "Who's the fastest average speed in the 1994 times with 1998 distances?"
7. Did some calculations/conversions.
8. Represented our data in a graph; drew conclusions—formed new questions.
9. Went back to 1998 times/data to more completely answer.

Figure 10.3.   Key steps in testing a conjecture using data.

- How expensive was it to set up for your sport at Nagano compared to other sports?
- How is scoring determined in your event? Give examples and discuss their "fairness" and objectivity.
- Within your chosen sport, how did the speeds of the 1998 Winter Olympic athletes compare in different events?
- How does "slope" affect speed in your sport?
- What are the slopes of the various events in your sport and how do they compare with slopes in other Winter Olympic sports?
- Has the number of countries participating and the number of athletes participating in your sport changed? Why do you think this is so?
- Who were the best athletes in your sport at the 1998 Winter Olympics? (Note that your answer to this question depends on how you decide to rank an athletes' performance, so you will need to write about what information you used to decide who were the best athletes and why you made this decision.)
- What country is best in your sport? (Note: I have information about the top 8 athletes in each event each year if you need it. Also, you will need to state what information you used to decide which country is best and why you made this decision.)
- How does U.S. performance in your sport compare with other Winter Olympic sports?
- How does U.S. performance in your sport this year compare with previous Winter Olympics?
- How do the ages of the 1998 U.S. athletes in your sport compare with other Winter Olympic sports? Weights? Heights?

Figure 10.4.   Some possible questions for students to
investigate about the Winter Olympics.

was kept in the classroom and we created folders of articles and results for the different Winter Olympic events that students could access as well (students were given extra credit points for bringing in resources that could be added to the class collection).

Other Teacher-Initiated Investigations

Throughout the whole-class investigation of the question on speed and the students' independent work on their research question, students were engaged simultaneously in four additional "compulsory exercises" (a term borrowed from the figure skating events). These activities were more teacher initiated, but were developed to pursue further students' interests while at the same time introduce some mathematical tools that I felt were important for them to learn. The four additional activities focused on the following issues: (a) investigating the question "How has your sport changed over time?", (b) examining distances, (c) investigating the question "What affects a country's performance in the Winter Olympics?", and (d) investigating ranking issues using final medal counts.

Investigating the question "How has your sport changed over time?" This investigation was developed as the first compulsory exercise and was given to help students become more familiar with their sport in a purposeful activity. Students were expected to gather specific data regarding some aspect of how their sport had changed over time, create a line graph, describe and interpret their graph—noting important trends and changes and possible reasons for these trends or changes—and to make a prediction about what they thought would happen in their event in the 1998 Winter Olympics, justifying this prediction based on the data they collected.

Examining distances. In this investigation, the Winter Olympic event, the biathlon, was used as a context for students to gain a sense of distances, both in terms of establishing benchmarks and working further with conversions from metric to English customary units. Students used maps of their town and their school to determine distances associated with the biathlon in order to help them establish visual benchmarks for meters and kilometers, and to gain a better sense of these units of measure. As a culminating activity to this component, students were given a compulsory exercise which asked them to determine distances in their own sport, similar to the activities that we had engaged in as a class with the distances associated with the biathlon. That is, students were asked to estimate first how far some distances in their sport would be using reference points in the classroom or on a map of the school or town. They were then asked to convert these distances into appropriate English customary units and mark these distances on an appropriate map using the given scale. Finally, they were asked to reflect on their original estimates, discussing how accurate they were and why this was the case.

Investigating the question "What affects a country's performance in the Winter Olympics?" Based on one of the initial questions posed to the students in this unit, I decided to address the question "What affects a country's performance in the Winter Olympics?" as a context for students to create and interpret scatter plots. Springing from a class discussion, students selected a variable that they thought might affect a country's performance that was easily quantifiable (e.g., past performance, population, gross domestic product) and a variable to consider a country's performance in the Olympics (e.g., total number of medals won in the last three Winter Olympics, number of medals won in 1998) and created scatter plots to investigate the relationship between these two variables. After I modeled this activity, the students completed a compulsory exercise independently.

Investigating ranking issues using final medal counts. As another experience for students to realize the sometimes subjective nature of mathematical data, after the conclusion of the Olympics, students engaged in an activity based on the reporting of the final rankings of the various countries. This activity was based on an article by Borasi (1989) in which she noted that different countries report different rankings based on different ordering criteria (e.g., in the U.S. media, the countries are ranked according to the total number of medals won as the primary criteria, while in the Italian media, the primary criterion is the number of gold medals won). Students engaged in a whole-class conversation about issues of ranking and fairness as they emerged through different ways of reporting results. Students explored different ranking possibilities through the use of a spreadsheet to perform calculations and to sort the countries using the new criteria.

Culminating Activities: Independent Research Project, Poster Project and Sharing of Results, Unit Portfolio and Reflection

Independent research project. As noted previously, one of the key components of this unit was the students' independent inquiry, in small groups, on a question of their own. The purpose of this independent research project was for students to make sense of a question of their choice, gather data, analyze this data, draw conclusions, and present their findings in a meaningful way. The emphasis was on deep exploration, composition of ideas, and creativity and clarity of the final product. While most of the project was done outside of class, I also set aside some class time throughout the unit when groups could work together on their project. Groups could use the resources available in the classroom that had been collected throughout the unit and they were encouraged to access the internet for up-to-the-minute results and information.

Poster projects and sharing of results. I felt it was important at the end of this unit for students to pull together some of their key findings about their sports and to share some of this information with each other. To this end, students worked with their partner(s) to create a poster highlighting their major findings about their sport using articles that they had read, and work that they had done on any of the compulsory exercises and/or their research projects. Each piece of evidence that the students included on their poster was to be accompanied by a brief description of its purpose (What was interesting/surprising in this information? Why was it included?). In order to encourage students to begin to reflect explicitly on the mathematics that they used, the final component of the posters was to respond to the question: "How did mathematics help you understand your sport better?" Students then formally presented their posters to the class while the audience recorded interesting/new things that they learned as well as new questions they thought of and shared these in brief discussions following each presentation. Students were also given an opportunity to look more closely at other groups' posters during one final day celebrating the culmination of the unit.

Unit portfolio and reflection. As in the other two inquiry units that the students had engaged in throughout the year, the final component of this unit was designed for students to reflect on their experiences in the unit in terms of what *activities* they engaged in, what they *learned*, and *why* this was important. This was particularly important in this unit as the mathematics used was not explicitly pointed out, but rather embedded in studying a current event – the Winter Olympics. I first engaged the class in a large group conversation reflecting on the unit as a whole, followed by students completing an independent reflection in which they were asked to comment on what they found interesting in the unit, what they felt they had learned, and what parts of the unit they had trouble with. I also asked students to select artifacts from their work that supported their descriptions and comment on the significance of each piece that they included.

ANALYZING STUDENTS' LEARNING

As noted previously, prior to implementing each of the inquiry units I had established goals for students' learning and developed a variety of assessment tools that would provide me with information about students' learning across these goals. The key assessment tools that were designed for the unit "Investigating the 1998 Winter Olympics" are recorded in Figure 10.5.

In order to analyze students' learning of the established mathematical goals for the unit, I created a table that included all of the unit goals and sub-goals and the assessment tools that were used to evaluate students' performance in relationship to each of these goals (see Table 10.1 for an example of this table for Goals 4 and 5).

> - Class discussions investigating the question "Who's the fastest?"; class discussions/warm-up on benchmarks in the classroom; class discussion during biathlon activity (audio-taped and field notes recorded immediately after class)
> - Homework: Reflecting on speed
> - Compulsory Exercise #1: How has your sport changed over time?
> - Compulsory Exercise #2: Examining distances in your sport
> - Compulsory Exercise #3: Examining speeds in your sport
> - Compulsory Exercise #4: What affects a country's performance in the Winter Olympics?
> - Independent research project
> - Unit portfolio/reflection
> - Final exam

Figure 10.5.   Key assessment tools for
"Investigating the 1998 Winter Olympics"

In creating the framework for the analysis of student learning on each of the mathematical content goals it was important to have a design that would allow me to be able to identify student understandings and/or misunderstandings with respect to each of the unit goals. Toward this end, I created a spreadsheet with cells representing each student's performance on each of the assessment tools identified to measure a particular sub-goal. To complete this analysis, I systematically read through all written work and/or listened to the daily audio tapes that I had identified to measure each sub-goal and recorded each student's performance on that particular task with a "0," "0.5," or "1" to indicate whether that student demonstrated "no understanding," "partial understanding," or "complete understanding" of that sub-goal on that particular task (see Table 10.2 for an example of a section of this spreadsheet). (Note that a blank cell represents "missing data." For example, Goal 5c—the use of a spreadsheet—was not required on Compulsory Exercise #3, so a blank cell means that that particular student did not *choose* to use a spreadsheet rather than attempted to use a spreadsheet and used it incorrectly, which would be noted by a 0 or .5 in the cell.) Thus, each student's performance on each identified task was used as evidence to determine whether the student met the sub-goal within each unit. This provided me with a quantitative look at the class as a whole according to each mathematical content goal/sub-goal within each unit. Through the recording of students' performance on the various assessment items designed to measure particular goals and sub-goals, I was able to look across particular students *or* particular goals and draw conclusions about students' learning on each of these goals.

Table 10.1   An example of mapping created between
unit goals and assessment tools.

4. Students formulate meaningful questions and collect, organize, display and analyze appropriate data to answer these questions (*"big idea"*). *Students:*	
a. Formulate a meaningful problem or question that can be explored using data (includes revising, refining question based on data available)	Research Question
b. select appropriate data to be collected to answer the question and justify those choices	Research Question
c. organize the data collected in a meaningful way	Research Question
d. make inferences and convincing arguments that are based on data analysis	Research Question
e. evaluate arguments that are based on data analysis	Research Question
5. Students construct, read, and interpret tables, charts, and graphs (*factual*) *Students:*	**Assessment Tool**
a. construct, read and interpret tables (review)	Compulsory Exercise #3 Research Question
b. construct, read and interpret bar and line graphs (review)	Compulsory Exercise #1 Research Question Final Exam: speed question #1
c. use a spreadsheet to create tables and perform calculations – optional	Compulsory Exercise #3 Research Question
d. read and interpret scatter plots	Compulsory Exercise #4 Final Exam: scatter plot question #2
e. construct scatter plots (using appropriate scales, etc.)	Compulsory Exercise #4 Final Exam: scatter plot question #1

Table 10.2   A section of a spreadsheet created
to record students' meeting of goals

Winter Olympics Unit	Lisa	Annie	Adam	Doug
Goal 4a *(Research Question)*	0.5	0.5	1.0	1.0
Goal 4b *(Research Question)*	1.0	1.0	1.0	1.0
Goal 4c *(Research Question)*	1.0	1.0	1.0	1.0
Goal 4d *(Research Question)*	1.0	1.0	1.0	1.0
Goal 4e *(Research Question)*				
Goal 5a *(Compulsory #3)*		1.0	1.0	1.0
Goal 5a *(Research Question)*	1.0	1.0	1.0	1.0
Goal 5b *(Compulsory #1)*		1.0	1.0	1.0
Goal 5b *(Research Question)*	1.0	1.0	1.0	1.0
Goal 5b *(Final: Speed Quest. #1)*	1.0	0.0	0.5	1.0
Goal 5c *(Compulsory #3)*				
Goal 5c *(Research Question)*	0.5	0.5	1.0	1.0
Goal 5d *(Compulsory #4)*	1.0	1.0	0.0	1.0
Goal 5d *(Final: Scatter plot #2)*	0.5	0.0	0.5	1.0
Goal 5e *(Compulsory #4)*	1.0	1.0	0.5	0.5
Goal 5e *(Final: Scatter plot #1)*	1.0	1.0	1.0	1.0
Individual Student Totals	10.5	11	11.5	13.5
Total possible for each student	12	14	14	14

Examples of Students' Independent Research Projects and My Analysis

To provide the reader with a sense of the type of analysis that went into creating this spreadsheet for each of the goals and sub-goals, I will discuss below an analysis of two pairs of students' performance on the mathematical content goal "Students formulate meaningful questions and collect, organize, display and analyze appropriate data to answer these questions (Goal 4)." This mathematical content goal and its sub-goals are recorded in Table 10.1. The primary assessment tool used to measure students' success on this goal was the research project. The guidelines given to the students for this research project can be found in Figure 10.6. Following each example is a brief analysis of how I feel the pair met the established goals for the

---

**PART II – ORIGINAL PROGRAM
MORE SPECIFICS**

The final product for this part of your project will be a report that includes tables, graphs and written descriptions summarizing your work and results. More specifically, this should include the following:

- The <u>original</u> question that you chose clearly stated. Also, how did your question <u>change</u>? That is, what is the more specific question that you answered based on your preliminary explorations and the data that was available to you. Describe how/why your question changed.
- A table containing the data that you collected/used to help answer your question. Then, write a paragraph describing what decisions you had to make in <u>choosing</u> your data. How/why did you make these decisions?
- A graph representing your data so that it is easy to understand. Be sure to carefully select an appropriate scale, label and title your graph(s).
- Your written conclusions based on the data you collected, your graph, and the background information that you found out in researching your sport. <u>This should be a rich description summarizing your work and you conclusions as well as the other research you have done. Include references to newspaper or magazine articles to support your arguments.</u>
- Your table(s), graph(s) and written work should be neat, organized, and easy to understand. Group work will also be included in your final grade.

Figure 10.6.   Independent research project description

project, in order to provide the reader with a better sense of the analysis that took place.

Annie and Lisa. The first pair, Annie and Lisa, began with the research question "What country is best in figure skating?" (see Figure 10.7). In the range of students' final products on this research project, I feel this one was in the middle—that is, there were some research projects that I would consider more complete and thorough in presentation and analysis than this one, and some that I would consider less. I believe Annie and Lisa's final product illustrates some of the difficulties that students had with this project, while at the same time illustrating some of the areas in which I felt students performed very well.

As described in their response to #1 in the written report, Annie and Lisa decided to determine which country was best in figure skating by

Country	# of Gold Medals	# of Silver Medals	# of Bronze Medals	# of Medals in 1998 Olympics	Total
Russia	5	5	3	5	18
USA	4	4	4	4	14
China	1	2	0	1	4
France	0	1	1	2	4
Canada	1	0	2	1	4
Germany	1	1	0	1	3
Finland	0	1	0	0	1
Czech Rep.	1	0	0	0	1

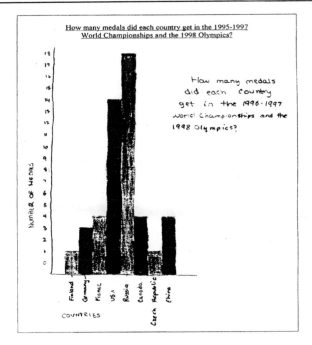

Figure 10.7.   Annie and Lisa's Olympic research project

determining the total number of medals that specific countries won in the 1995–1997 World Championships and the 1998 Winter Olympics for all figure skating events. However, although Annie and Lisa did select criteria to determine "best" and used appropriate data to answer their question, they did not justify why or how they made this decision about their data (noted by 0.5 in the spreadsheet for Goal 4).

Annie and Lisa did organize their data appropriately in a table and a bar graph to present which country won the most medals in the 1995–1997 World Championships and the 1998 Winter Olympics. Although this was an appropriate choice for a graph, in their description of the graph the girls wrote, "Our graph is a bar graph which represents each country and the number of gold, silver, and bronze medals in the World Championships compared to how many were won in the 1998 Winter Olympics," which was *not* what the bar graph illustrated. The title for their bar graph was correct, however, so this discrepancy may have been a result of miscommunication between the two girls—a common issue in the groups—since I know that Annie typed up the explanation in the report and Lisa created the bar graph. Annie and Lisa also came in to the computer room one afternoon to set up a spreadsheet of their data. However, the spreadsheet was used simply for a table, and not to perform any calculations (noted by 0.5 for goal 5c).

I was disappointed that Annie and Lisa did not make a final conclusion in their written report as to which country they felt was the "best" in figure skating. However, on the poster that they created summarizing their sport, Annie and Lisa included the table and bar graph from this independent research project. They summarized their findings on this poster: "We now know, by gathering information for our table and graph, that Russia has the highest number of medals in the 1998 Olympics and the World Championships in 1995, 1996, and 1997...Just because Russia has the highest number of medals in figure skating, that doesn't mean that they are the best figure skating country. They might just have a lot of medals because they have a lot of talented figure skaters. Finland or Canada might have had a better figure skater, but those countries did not have enough figure skaters to beat Russia's number of medals." This illustrates an interpretation of their findings in their data and a realization that their determination of "best" was dependent on their criteria of total number of medals won in those events/years.

Adam and Doug. The second pair, Adam and Doug, chose to investigate the question "How does the luge speed compare with the speed of other Olympic events?" (see Figure 10.8). These two students were more successful on their project than Annie and Lisa were, particularly in formulating a revised question based on the data available and then in more carefully answering this question.

Adam and Doug chose to re-visit the idea of speed that had been introduced earlier in the unit to consider how the average speed of the luge compared to the average speed of some other Winter Olympic sports. They did carefully re-state their question, noting that they were finding "average speed" and they discussed some decisions that they made about their data (e.g., what events to include in their analysis and from what years, what units to use). They also accurately provided a table with their data converted from meters per second to miles per hour, a corresponding bar graph of the data, and a written description of their results and analysis.

As can be seen, Adam and Doug's research project included all of the required components of the project (as listed in Figure 10.8) and their table, graph and written report were consistent in supporting/answering their revised question about the average speed of various events in the 1998 Winter Olympics. One of the things I would have liked this pair to comment on, however, particularly in light of our discussions about speed earlier in the unit, is that a given sport may have a faster *average speed* for the entire race, but not have the fastest *instantaneous speed.* There would certainly not have been any way for this group to have found these instantaneous speeds, but throughout the television viewing of these events, these speeds were often reported. They did use the data available to them, however, to answer their question.

This independent research project served to assess a number of goals—both in terms of assessing students' mathematical content knowledge as well as their ability to engage in a research process. In terms of mathematical content goals, I was able to assess students' ability to construct and interpret tables and graphs, as well as some students' ability to use a spreadsheet to create tables and perform calculations. The key component, however, was students' ability to formulate meaningful questions and collect, organize, display and analyze appropriate data to answer these questions. These are certainly not straightforward tasks in and of themselves, and require higher-level thinking skills on the students' part. This was particularly the case with this project because of the complex nature of posing questions and gathering data in a real-life context. This was a component of one of my underlying goals in all of the inquiry units–for students to begin to view mathematics not as a pre-established body of knowledge, but as a tool that can be used in many different ways and is very dependent on and reliant on choices/decisions.

These two pieces of student work illustrate this complexity. For example, the idea that asking what one may consider to be a simple question—"What country is best in figure skating?"—can actually be answered in

Our original question for this project was: how does the luge speed compare with the speed of the other events. After finding difficulty in gathering the data we needed to do this, we decided to only compare the luge to five other Olympic speed events: the four man Bobsled, the men's 10000 and 5000 meter speed skating events, the men's Super G, and the men's Giant Slalom. We based our speed ratings on a miles per hour basis. We also decided to use the results from the 1998 Winter Olympics in Nagano, Japan, because we were unable to find course lengths from past Winter Olympics. After all of the changes that we made due to lack of information, our original question changed and became much more descriptive. Our final question was: how does the average speed in miles per hour of the gold medal time in the 1998 Winter Olympics of the two man luge compare to those of five other Winter Olympic speed events?

There were many small decisions that we had to make while researching and writing our project. Because we were unable to find course lengths from past Olympics we were required to research only the Nagano Olympics. When it came to measuring speed we had to decide whether to measure in miles per hour or meters per second. We decided to measure in miles per hour because the scale of miles per hour is much more familiar to us because we do not use the metric system in the U.S. Along the way we also encountered the problem of time. Because there are so many different Winter Olympic events we decided to only calculate five other events in addition to the two man Luge. We picked five events that we thought demonstrated a wide variety of speeds and would make our project more interesting. To gather our information we used sources such as the CBS Winter Olympic web site, the information packets that you handed out at the beginning of the Olympic unit, and the course size information and event results posted on the walls of the classroom.

After graphing, charting and analyzing all of the data that we collected we found that the two man Luge was the third fastest event of the six events that we charted. The average speed in miles per hour of the gold medalist at Nagano came out to be 52.80564999 miles per hour. Like many events at the Winter Olympics the luge is a sport based upon speed. The person with the shortest cumulative time after two runs, wins. We cannot say for sure that the luge is the third fastest event, because we did not look into many of the other Winter Olympic sports. After also graphing the times that it took the lugers to complete the Olympic courses over the past years it appeared that the times had become slower and slower. The only explanation for this would obviously be that the courses have become longer and longer over the years. Our calculations for the average speed of the gold medal two man Luge came out to be 52.80564999 mph. Other records have shown the Luge to go much faster than this. The only logical explanation for this would be that the Luge course that these readings were taken on happened to be a steeper course.

In the two man Luge, two men ride down a narrow lane of solid ice on a small lightweight sled. The bottom man lies flat on the sled and holds on to two straps attached to the top man's gloves. The top man lies on top of the bottom man and steers the sled with the sides of his legs. If the pair crashes on the course they are allowed to get back up and finish the course on their sled. The Luge is considered one of the most dangerous Olympic sports. The sledders lie only 3 inches off of the ice. Hitting solid ice at speeds above 50 miles per hour can be deadly.

Events Included	Ranking (fastest to slowest)	Average Speed in mph	Distance in meters	Time in Seconds
Men's 10000 Meter Speed skating	1. Four Man Bobsled	57.25565093	1360 (3 runs)	159.41
Men's Giant Slalom	2. Men's Super G	57.16460771	2423	94.82
Four Man Bobsled	**3. Luge – Men's Doubles**	**52.80564999**	**1194 (2 runs)**	**101.105**
Men's Super G	Men's Giant Slalom	41.97184432	1487 (2 runs)	158.51
Men's 5000 Meter Speed skating	Men's 5000 Meter Speed skating	29.26530631	5000	382.2
Luge – Men's Doubles	Men's 10000 Meter Speed skating	28.12719249	10000	795.33

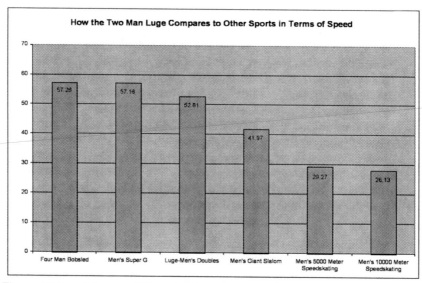

Figure 10.8   Adam and Doug's Olympic research project.

many different ways depending on what data one chooses to collect and how one decides to analyze and represent that data. In other words, one of my goals was actually to problematize for the students asking questions, collecting, and analyzing data. These projects, then, can be viewed with not simply the lens of *collecting, analyzing and representing data* but also the lenses of *what issues/difficulties/decisions arose* as students went to investigate their question.

Considering the class's performance as a whole on Goal 4, I was very impressed with students' level of involvement and performance in this difficult process. All eight groups formed a question to explore using data and were able to make inferences and convincing arguments based on their data analysis. Some of the groups had difficulty selecting appropriate data to collect and to justify the choices made in data selection (e.g., Annie and Lisa). Two groups also did not answer explicitly the question that they posed. This was not surprising given that drawing conclusions based on data analysis was not an easy or straightforward task, and required students to consider critically their data and what information it provided in terms of the question posed. Through these examples and analysis, one can gain an overall sense of how the spreadsheet created for each goal was used to look at data across a particular goal or across a particular student.

MY OVERALL FINDINGS

In conclusion, in performing this research in my classroom there were a number of things that were highlighted for me as both the teacher and the researcher. These learnings are described below.

Establishing and Measuring Instructional Goals is Much More Complex and Challenging Than It May Seem

The systematic process that I engaged in *as a researcher* revealed for me *as a teacher* the complexities involved in what and how we assess students' learning. Based on my experiences, middle school mathematics teachers typically design assessment questions for summative assessments to measure how students can perform on material that the teacher has presented them with in class (and usually practiced on homework). Teachers then grade these questions, rating the students on whether they could do the problems, without actually assessing or reflecting on what it is that each student knows and does not know. In contrast, in designing the assessments for the inquiry units I wanted to provide students with multiple opportunities to show what they knew and could do throughout the unit, often asking them to engage in assessment tasks that required them to apply their mathematical understandings to new contexts or situations. Designing these tasks is not easy and it calls for many of us to re-think what it means to grade/assess students' learning.

Different Forms of Assessment May Provide Different Glimpses Into Students' Understandings

This study made very clear to me that using several assessment tools to measure a particular goal may be critical to enable *all* students to demonstrate what they have learned. For example, how students performed in class discussions was often not consistent with how they performed on a written task on the same concept. One student in particular, actively participated in classroom discussions, often offering excellent mathematical insights, but did not produce consistent results on written tasks, particularly on the final exam. On the other hand, another student rarely, if ever, participated in classroom conversations, but often had excellent ideas recorded in writing. This provides support to the idea that teachers need to provide students with multiple opportunities to show what they know and can do, and they also need to provide feedback in order to help students to continue to grow in their learning.

Students May Not Demonstrate Consistently What They Learned

The analysis of student learning was more difficult than I had anticipated it would be. I thought that measuring students' learning of the content goals in particular was going to be rather straightforward—I would consider the assessment tools designed to measure that goal and would be able to draw easily some conclusions about what the students knew. However, this was not usually the case. One of the main reasons was that, more often than not, students did not perform consistently across assessment tools designed to measure the same goal. That is, it often happened that a student would demonstrate understanding of a certain concept on the first assessment tool considered, and then on the next would not, but perhaps on a third one again s/he would demonstrate understanding. Therefore, there was not always a straightforward progression from "not understanding" to "understanding." I often found it very challenging, then, to draw conclusions based on this data. This also forced me to think more carefully about the assessment tools used and what components may have helped or hindered students from demonstrating their understanding. Also, as the goals themselves were often complex, so were the assessment tools used to measure these goals, as demonstrated by the analysis of the two research projects.

CONCLUSIONS

Due to the complexity involved in measuring the instructional goals for school mathematics set by the NCTM *Standards* (1989, 1991, 1995, 2000), and the time it took me to conduct my analysis of the data collected through the various assessment tools I employed, it is probably unrealistic to assume that teachers will be able to carefully analyze and report about students' learning in the same depth on an ongoing basis. Yet, this is an extremely important component of the recent reform movement in mathematics—that is, providing evidence of the mathematics that students can accomplish when taught mathematics based on reform movement recommendations. It is critical to be able to share with others what students know and can do as those involved in mathematics education grapple with how to best teach our students.

It is also important to note that of the six goals identified for the inquiry unit described here (see Figure 10.1), only three of those goals have typically been a focus of middle school mathematics: (a) estimate and performing measurements (Goal 3), (b) construct, read and interpret tables, charts and graphs (Goal 5), and (c) calculate and compare measures of central tendency (Goal 6). Through systematic documentation and analysis, this study provides evidence that students can move beyond the learning of those more factual goals and engage in the learning of big mathematical ideas such as formulating meaningful questions, collecting, organizing, displaying and analyzing appropriate data to answer these questions (Goal 4). Thus, this study contributes to the field of mathematics education by providing a rich image of theory in practice combined with a systematic analysis of students' learning in the process.

These results have confirmed and further elaborated my initial and more intuitive evaluation as a teacher of the potential of an inquiry approach to enhance students' learning outcomes in mathematics. I hope that these findings will provide the necessary motivation to encourage more mathematics teachers to make the efforts necessary to move in the direction indicated by the most recent calls for school mathematics reform, as articulated by the NCTM's (2000) *Principles and Standards for School Mathematics.*

REFERENCES

Borasi, R. (1992). *Learning mathematics through inquiry.* Portsmouth, NH: Heinemann.

Borasi, R. (1996). *Reconceiving mathematics instruction: A focus on errors.* Norwood, NJ: Ablex,

Callard, C. (2000). *An in-depth look at Students' learning in an eighth grade mathematics classroom informed by an inquiry approach.* (Doctoral dissertation, University of Rochester, 2000).

Cobb, P., Wood, T., & Yackel, E. (1991). A constructivist approach to second grade mathematics. In E. von Glasersfeld (Ed.), *Radical constructivism in mathematics education* (pp. 157-176). Dordrecht, The Netherlands: Kluwer.

Confrey, J. (1990). What constructivism implies for teaching. In R. B Davis, C. A. Maher, & N. Noddings (Eds.), *Constructivist views on the teaching and learning of mathematics* (pp. 107-122). Reston, VA: National Council of Teachers of Mathematics.

Davis, R., Maher C., & Noddings, N. (1990). Constructivist views on the teaching and learning of mathematics. In Davis, R., Maher, C. A., & Nodding, N. (Eds), *Constructivist views on the teaching and learning of mathematics* (pp. 1-3). Reston, VA: National Council of Teachers of Mathematics.

National Council of Teachers of Mathematics. (1989). *Curriculum and evaluation standards for school mathematics.* Reston, VA: National Council of Teachers of Mathematics.

National Council of Teachers of Mathematics. (1990). *Professional standards for teaching mathematics.* Reston, VA: National Council of Teachers of Mathematics.

National Council of Teachers of Mathematics. (1995). *Assessment standards for school mathematics.* Reston, VA: National Council of Teachers of Mathematics.

National Council of Teachers of Mathematics. (2000). *Principles and standards for school mathematics.* Reston, VA: National Council of Teachers of Mathematics.

National Research Council. (1989). *Everybody counts: A report to the nation on the future of mathematics education.* Washington, DC: National Academic Press.

New York State Department of Education. (1994). *Curriculum, instruction, and assessment framework for mathematics, science, and technology.* Albany, NY: New York State Department of Education.

New York State Department of Education. (1996). *Learning standards for mathematics, science, and technology.* Albany, NY: New York State Department of Education.

Simon, M. (1995). Reconstructing mathematics pedagogy from a constructivist perspective. *Journal for Research in Mathematics Education, 26,* 114-145.

Webb, N. (1992). Assessment of students' knowledge of mathematics: Steps toward a theory." In D. Grouws (ed.), *Handbook of research on mathematics teaching and learning* (pp. 661-683). New York: Macmillan.

Woodward, A., Borasi, R., & Packman, D. (1991). National Science Foundation Grant TPE 9153812 Proposal: *Supporting middle school learning disabled students in the mainstream mathematics classroom.* Rochester, NY: University of Rochester.

CHAPTER 11

RELATING CLASSROOM INTERACTION TO STUDENT ASSESSMENT RESULTS

Mary C. Shafer
Northern Illinois University, De Kalb, IL

Annette Hill
South Dade Senior High School, Homestead, FL

A teacher's classroom assessment practice supports student learning of significant mathematics and provides information for instructional decision making. As such, assessment is an integral part of instruction (NCTM, 2000). For many teachers, integrating teaching and assessment practices is difficult to achieve, and examining long-term results of such efforts is seldom considered. The purpose of this chapter is twofold: (a) to present analysis of the first author, Mary, with respect to the nature of the teaching and assessment practices of the second author, Annette, and (b) to discuss results from constructed-responses assessment items from students in her classes. Through this process, the authors investigate dimensions of classroom interaction that supported student understanding of mathematics as

Teachers Engaged in Research
Inquiry Into Mathematics Classrooms, Grades 6–8, pages 219–246

measured by the assessment items. The chapter closes with implications of this work for teachers and educational researchers.

THE SETTING

Mary and Annette were both involved in research funded by the National Science Foundation to investigate the impact of *Mathematics in Context* [MiC] (National Center for Research in Mathematical Sciences Education & Freudenthal Institute, 1997–1998) on student achievement. MiC is a Grades 5–8 curriculum designed to address the National Council of Teachers of Mathematics (NCTM) *Curriculum and Evaluation Standards* (1989); its development was supported by the National Science Foundation. MiC provides opportunities for students to deepen their understanding of significant mathematics in number, algebra, geometry, probability and statistics while emphasizing connections among mathematical ideas. Ample opportunities are provided for students to solve experientially real problems designed to stimulate mathematical thinking. One of the purposes of the MiC longitudinal study was to evaluate the impact of the MiC instructional approach by tracking changes in students' understanding of mathematics over time and by examining relationships between students' performance and their preceding achievement, the instruction they experienced, their opportunity to learn mathematics with understanding, and the capacity of their schools to encourage and sustain high academic expectations (Romberg & Shafer, 2004). Beginning in the 1997–1998 school year, data were gathered over a three-year period in four school districts; 90 teachers from 19 schools and over 2200 students participated in the study.

As research coordinator for the MiC longitudinal study, Mary worked with various school districts to identify teachers to participate in the study. Annette was recommended by her principal to be involved as a study teacher. He believed that she was a very effective middle school mathematics teacher and frequently chose her to be a representative of the school for official guests who wanted to see high-quality classroom instruction in action. Annette participated in the study as a seventh-grade teacher during the 1997–1998 school year and as an eighth-grade teacher the following year. She characterized MiC as:

> a very different approach to teaching mathematics. It forces teachers to think of mathematics differently than in prior years. It is extremely different from the way we were taught. The approach is innovative, interactive, rich in context, and hands-on, but it requires much preparation to teach.

Annette felt that the MiC longitudinal study supported her efforts to implement a new curriculum for the first time. The middle school in which she taught is located in a large, urban area in the southeastern United States. More than 60% of the nearly 1300 students enrolled in the school were minority students—32% were African American students and 32% were Hispanic students. More than 50% of the students in the school were eligible for government-funded lunch programs.

Mary worked with Annette during professional development institutes sponsored by the study and in coordinating study participation in her school. Mary and Annette have extensive experience teaching mathematics—10 years and 23 years at the middle school level, and 6 years and 2 years at the high school level, respectively. During her participation in the study, Annette completed a master's degree in mathematics education. She currently teaches mathematics at the high school level and is chair of the mathematics department at her school. Mary is currently completing the analysis of data from the longitudinal study. In this role, Mary analyzes classroom instruction, classroom assessment practice, and students' opportunity to learn for all study teachers, and assessment results for students who participated in the study.

During the fall of 2003, Mary and Annette decided to collaborate in a smaller study that had two purposes. First, we reviewed assessment results for a group of students from Annette's classes who had completed assessments administered during the longitudinal study over a two-year period. Our purpose was to learn about the perspectives a teacher would bring to and develop as she reviewed responses from her own students to constructed-response items that assessed both mathematical reasoning and skills. Second, we wanted to examine these results with respect to Annette's teaching and assessment practices. Our goal was to use Mary's analyses of these practices, based on observations conducted in Annette's classes during the longitudinal study, to infer connections between student responses and the instruction students experienced in Annette's classroom. The results of this research are presented in this chapter.

Student responses to assessments designed for the longitudinal study were scored during multiple scoring institutes. Some were scored during institutes held at each research site directly after the tasks were administered to the students. Assessments were marked with a numeric code for anonymity and shuffled before scoring. Teachers, therefore, scored assessments from a variety of study teachers; they had no way of determining whether the responses they scored were from their own students. The collaboration between Mary and Annette was the first time that a study teacher was able to review work from students in her own classes. Copies of actual student responses were used in this collaboration, which allowed for consideration of the entire response workspace. Since students were in

Annette's classes several years prior to this collaboration, recognition of student handwriting was not an issue. In discussion of student responses in this chapter, pronouns such as his and her are used in a generic way and are not meant to indicate a student's gender.

TEACHING FOR UNDERSTANDING

Research in a growing number of studies underscores the importance of teaching mathematics for understanding. Teaching mathematics for understanding is based on the principles that knowledge is constructed by the learner and is shaped by the learner's existing knowledge, skills, and beliefs; the teacher's role is a guide for facilitating conceptual understanding; mathematical tasks engage students' thinking about important mathematics; and classrooms are communities of learners (Cohen, McLaughlin, & Talbert, 1993; Fennema & Romberg, 1999; Hiebert et al., 1997). When learning mathematics with understanding, students need the time and opportunity to develop relationships among mathematical ideas, extend and apply these ideas in new situations, reflect on and articulate their thinking, and make mathematical knowledge their own (Carpenter & Lehrer, 1999).

The tasks used during instruction convey messages about the nature of mathematical activity and shape students' opportunities to learn mathematics with understanding. An important role of the teacher is sequencing these tasks in ways that provide opportunities for reflection and discussion. Instruction is enhanced when students are allowed to do the mathematical work involved in the tasks (Stigler & Hiebert, 1999). Tasks, however, can be adjusted to either encourage or reduce cognitive activity. The difference resides in teachers "proactively and consistently support[ing] students' cognitive activity" (Henningsen & Stein, 1997, p. 546).

Classroom discourse, including expressing thinking and representing mathematical ideas, is central to inquiry-based mathematics learning (NCTM, 1991). Substantive discourse promotes shared understanding of mathematical ideas and emphasizes higher order thinking, which requires students to "combine facts and ideas in order to synthesize, generalize, explain, hypothesize or arrive at some conclusion or interpretation" (Newmann, Secada, & Wehlage, 1995, pp. 86–87). Such conversation is characterized by reciprocity. Students listen carefully to each other's ideas, build conversation on those ideas, and mutually construct their understanding.

Teachers seek evidence to confirm their intuition regarding students' mathematical skills, disposition, knowledge, and understanding. They gather evidence, for example, to evaluate whether students possess the prerequisite knowledge for completing a task, correctly interpret the parameters of the task, and are ready to engage in tasks on their own. For valid evidence to be gathered, the learning environment must be non-threatening and conducive to sharing and valuing others' perspectives. Effective class-room assessment practice requires a wide range of evidence that attends to both process (e.g., reasoning and communication) and product (solutions). Teachers who exhibit exemplary classroom practice create and seek opportunities to assess student learning beyond quizzes and tests, recognizing that multiple sources of evidence are embedded in instruction (Shafer & Romberg, 1999).

DESCRIPTION OF ANNETTE'S TEACHING AND ASSESSMENT PRACTICES

Using data from classroom observations, journal entries, and interviews gathered during the MiC longitudinal study, we describe Annette's teaching and assessment practices through a vignette based on a portion of one observed lesson. Throughout this description, we identify key elements that promote mathematical understanding.

Since the class met every other day, Annette began each class with a review of the lesson from the previous class period. On this day, the review was related to the "Beams" lesson in *Building Formulas* (Wijers et al., 1997, pp. 11–13). In this lesson, students completed a table to organize the number of rods in beams of different lengths. For example, the beam of length three shown in Figure 11.1 is composed of 11 rods. Annette inquired, "What did you learn in the Beams lesson during the last class period? Thomas, what is the number of rods for a beam of length two?" When he responded six rods, Annette pushed him to defend his answer: "How did you come to that conclusion? Come to the board and show us how you determined six rods for a length of two." As Thomas counted the rods, he realized his mistake. Annette noted:

> As I continued questioning, I noticed several students erasing their answers as they recognized errors in figuring the total number of rods. Other students traced the figures in the unit with their pencils, or put tick marks on the rods in each figure, trying to come up with a way to accurately find the number of rods.

The length of the beam is the number of rods along the underside.

Figure 11.1.   The beams context from Building Formulas (Wijers, et al., 1997, p. 11).

At this point, she drew the table on the board that was a variation of the table in the lesson (see Table 11.1) and asked the students to recalculate the totals and justify their solutions. She supported their thinking by encouraging visualization: "When you can visualize the number of rods in a beam, then record the number in the table. I want you to explain how you determined each number." Annette directed students' attention to important elements of the task, but she did not reduce the cognitive challenge by completing the table for them. Rather, she pressed them to visualize the pattern and to explain each entry in the table. This process created a basis for students to notice the relationship between the number of rods in consecutive beam lengths, a critical element in developing a recursive formula of the pattern.

Annette used student ideas to work toward shared understanding for the class. For example, Jose mentioned that he used a shortcut to determine the number of rods in a beam. Rather than counting each rod, Jose just added the additional rods he drew for the next beam to the total number of rods for the previous beam. Jose explained, "If the total number of rods for Beam 2 is seven, I add four more rods to the picture to get Beam 3. So 11 is the total number of rods for Beam 3." Jose went to the board to demonstrate what adding four rods would look like on the diagram. Starting at the end of Beam 2, Jose drew another triangle and connected the top of the previous beam to the top vertex of the new triangle. Annette used Jose's method to spark a discussion of other alternative strategies. These strategies helped students in developing both recursive and direct formulas.

Table 11.1.   Table Annette used during class review of the Beams context from Building Formulas (Wijers et al., 1997, p. 10)

Length of the Beam	Drawing of the Beam	Total Number of Rods in the Beam
1		
2		
3		
4		

After completing the table for beams up to length four, Annette asked students to conjecture about the number of rods in a beam of length eight. She encouraged students to make drawings to help them investigate their conjectures. Annette continually supported higher-order thinking by posing questions such as: How did you visualize a beam of length eight? What did you think about first? Why? How does your drawing fit the pattern? How can you verify the number of rods in the beam? In doing so, Annette promoted connections between the geometric design and the table of values. She pressed students to communicate effectively their thinking, not just the steps they used to determine their answers. Annette also used questioning techniques to help students refine explanations. During group work, for example, Shawna expressed the recursive relationship for the pattern: "If you know the length of a beam, just add four to get the length of the next beam. That's the pattern I see in the numbers in the table." Annette queried, "How do you know that's true, Shawna? Use the drawings to show how that works. Then write that as part of your explanation." In this way, Annette provided guidance, but students completed the mathematical work on their own.

In the next part of the class period, Annette led a class discussion of the homework, which was the first two pages of the "Bricks" lesson (Wijers et al., 1997, pp. 17–18). In the lesson, students described the basic or core patterns in three different rows of lying and standing bricks using words and symbols (see Figure 11.2). Students lengthened the brick patterns by repeating the basic pattern, and they discussed how to find the number of bricks in any row by using the basic pattern. This discussion was the foundation for the new MiC classwork assignment. Students worked in small groups as they drew brick patterns, determined the length of particular brick patterns, and evaluated both recursive and direct formulas for determining pattern length. As students worked on these pages, Annette posed questions such as: What information are you given to help you answer this question? What do you need to know to find a solution? What do you think that term in the formula means? How does this formula relate to the drawing of the bricks? How does the new formula relate to the previous one? She encouraged every student to respond to one of her questions during group work and justify the accuracy of the response. Students worked predominately individually rather than as a team; group members had not fully supported and taken responsibility for each other's progress. Yet when students did have questions, they readily asked other group members for assistance.

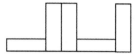

This basic brick pattern can be symbolized as LSSLS, where L represents a lying brick and S represents a standing brick.

Figure 11.2.   A basic brick pattern students used in
Building Formulas (Wijers, et al., 1997, p. 18).

When this occurred, their conversations were often substantive, discussing strategies for solving problems. For example, Christina asked Thomas to help her write an expression for a length of bricks using conventional symbols in place of LLSSSLLSSSLLSSS. Thomas asked her to circle the basic pattern each time it showed up in the letters. After Christina circled "LLSSS" three times, he guided her in writing 2L + 3S for the basic pattern, and 3 x (2L + 3S) for the entire length. Students were very much on task during the whole lesson, eager to contribute to large group discussion, and productive as they worked in groups. Throughout the class period, students experienced multiple opportunities to reason through mathematical tasks in both MiC lessons with support of Annette's probing questions and the exchange of ideas by class members.

Annette's teaching and assessment practices were well integrated. For example, Annette noted at the beginning of the class period that she recognized students' errors in determining the number of rods in particular beams. In response to this evidence, Annette drew a table on the board, asked students to reevaluate their initial solutions, and asked students to explain subsequent solutions. Annette noted: "If teachers listen to and observe their students, they will reflect on their students' thinking. This will get teachers to think about what students are actually doing. It changed my thinking." Annette's assessment practice allowed her to identify student difficulties, and the information she gathered led to instructional decisions that promoted conceptual understanding of the mathematics.

STUDENT ASSESSMENTS

In this section, we examine student responses to assessments, and we investigate the relationship between Annette's practices and student responses. Annette "looped" with her classes—that is, she moved with her students from seventh to eighth grade. Since she taught the same students in both grades, we explore how Annette's teaching and assessment practices influenced the quality of student responses over time.

Student Sample

We reviewed assessments from the 14 students who took both assessments designed for the MiC longitudinal study (described below) in both seventh and eighth grades. Four identified themselves as African American students, three as Hispanic students, one indicated that he was multiracial, and the rest were Caucasian students; eight students were girls. Students in the sample had similar mathematical problem solving profiles at the beginning of the seventh-grade. On the traditional portion of the state standardized test, national percentile rankings in mathematics application for the sample clustered in the average range at the beginning of the study. Less variation in these scores was evident from the beginning of seventh grade to the end of eighth grade. Students scored higher in algebraic thinking than in other categories on the performance-based portion of the state testing system administered in the eighth grade.

Description of Assessments

In the analysis, we used student responses from the two assessment systems developed for the MiC longitudinal study. The External Assessment System [EAS] (Romberg & Webb, 1997–1998) is a set of grade-specific assessments designed to measure student performance on publicly-released tasks that were used in national and international tests. In order to examine growth over time, a set of items of moderate difficulty, called anchor items, were repeated on each grade-specific assessment. The remaining non-anchor items increased in relative difficulty from one grade to another. Twenty percent of the items were constructed response. Responses were scored with the rubrics used on the national and international tests.

To assess broader curricular goals, a Problem Solving Assessment System [PSA] (Dekker et al., 1997–1998) was developed to address three levels of reasoning: (a) conceptual and procedural knowledge, (b) making connections, and (c) mathematical analysis and generalization. All items were set in contexts and were constructed response. A series of algebra items across the set of grade-specific assessments was designed to assess the development of algebraic reasoning over time. Items were generally situated in the same context and were increasingly difficult over the grade levels. Full or partial credit was awarded based on accuracy of response and thoroughness of explanation.

We chose to focus this analysis on algebra items for three reasons. First, rather than teaching a series of procedures, the algebra units in MiC build on student reasoning. This emphasis provides multiple ways for students to solve problems, use representations, formulate generalizations, and communicate reasoning. Second, some algebra items on each assessment were designed for examining the development of algebraic reasoning over time. Third, Annette expected that students would perform better on algebra items than items from other content strands. Students had worked at generalizing patterns and had studied several MiC algebra units over the two-year period that she and her students participated in the longitudinal study.

Results of External Assessments

Two constructed-response algebra items were included in the EAS. The first item, Similar Triangles, a publicly-released item from the Third International Mathematics and Science Study [TIMSS] (International Association for the Evaluation of Educational Achievement, 1996), was included on both the seventh- and eighth-grade EAS (see Figure 11.3).

> This publicly-released item from the Third International Mathematics and Science Study (TIMSS; International Association for the Evaluation of Educational Achievement, 1996) appears in the External Assessment System (Romberg & Webb, 1997–1998). Students interpret a pattern demonstrated in diagrams and extend the pattern to the 8th figure.
>
> Here is a sequence of three similar triangles. All of the small triangles are congruent.
>
> The sequence of similar triangles is extended to the 8th Figure. How many small triangles would be needed for Figure 8?

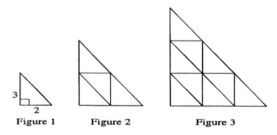

Figure 1   Figure 2   Figure 3

Figure 11.3.   "Similar Triangles" algebra assessment item in the External Assessment System.

When reviewing student work, we examined each student's responses in seventh and eighth grades together in order to look for differences over time. In the task, students examined a pattern of triangles in three figures and extended the pattern to find the number of triangles in the eighth figure. As seventh graders, three students answered correctly, although all students attempted an answer (see Table 11.2). As eighth graders, two more students responded correctly.

Although only a few students in our sample received full credit, through looking at actual student work, Annette was able to see some progress in the ways students approached the task:

In seventh grade, two students stopped at drawing the fourth pattern number, then generalized the pattern (n^2) and used it to find the correct number of triangles in the eighth pattern. In eighth grade, another student used both n^2 and a number pattern to find the result. Other students who responded incorrectly the first year, like those who used an inaccurate drawing strategy, used an algebraic expression the second year. One student, who did not show any reasoning in the first year, knew to look for some kind of comparison from one pattern number to the next in the second year. Clearly, I see that my kids did show progress on this task from year to year. Even though some students did not understand the task in the first year, they knew how to look at the problem in the second year.

The second constructed-response algebra item, Marci's Dots, a publicly-released item from the National Assessment of Educational Progress

Table 11.2.   Results of algebra constructed-response
items from the External Assessment System

	Full Credit*	Partial Credit	Attempt	Non-Response
Grade 7				
Similar Triangles	3	0	10	1
Grade 8				
Similar Triangles	5	0	8	1
Marci's Dots	3	5	4	2
*Number of Students				

[NAEP] (National Center for Education Statistics, 1992), was a more challenging item used only on the eighth-grade EAS (see Figure 11.4).

In this task, students examined a pattern of dots in rectangular arrays, extended the pattern to the twentieth term, generalized the pattern, and wrote an explanation of their reasoning. Three students were awarded full credit, and five students attempted to draw the sequence. In reviewing student work, Annette was impressed with several responses:

This student used algebraic notation learned in a MiC algebra unit: Step n is $n \times n + 1$. Good job, although I'd like to see how he determined that worked.

Another showed a number pattern: $1 \times 1 + 1$, $2 \times 2 + 2$, $3 \times 3 + 3$, $20 \times 20 + 20$, to find the correct number of dots. This student created a table and completed it with correct values. She saw a pattern right away and explained it well (see Figure 11.5a).

This publicly-released item from the National Assessment of Educational Progress (National Center for Education Statistics, 1992) appears in the External Assessment System (Romberg & Webb, 1997–1998). Students interpret a pattern demonstrated in diagrams, generalize the pattern to determine the 20th term, and explain their process.

A pattern of dots is shown below. At each step, more dots are added to the pattern. The number of dots added at each step is more than the number added in the previous step. The pattern continues infinitely.

Step 1   Step 2   Step 3

2 dots   6 dots   12 dots

Marcy has to determine the number of dots in the 20th step, but she does not want to draw all 20 pictures and then count the dots. Explain or show how she could do this <u>and</u> give the answer that Marcy should get for the number of dots.

Figure 11.4.   "Marci's Dots" algebra assessment
item in the External Assessment System

Another student knew how to find the differences of differences, a method used in a MiC algebra unit (see Figure 11.5b). Good start. Degree in this case means an exponent of two, not to multiply by two. He knew two was the number of times it took to get to the constant of change. But he multiplied by two instead of raising it to the power of two.

Annette commented:

Overall, their understanding of algebra was good. They did well. Some students with partial credit showed some thinking to come up

(a) She has to look at the pattern and figure what's next. Here you have to multiply the previous step # by the next one. For example, 1st step # x 2nd step # = 2 dots, 2nd step # x 3rd step # = 6 dots, and so on. The 20th # is 420.

1	2	3	4	5	6	7	8	9	10	11	12	13	14	15	16	17	18	19	20
2	6	12	20	30	42	56	71	90	110	132	156	182	210	240	272	306	342	389	420

(b) Make a chart.

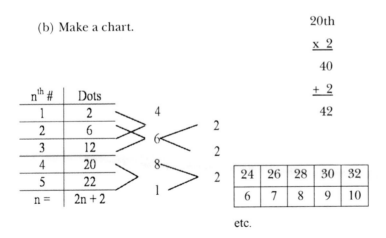

Figure 11.5.   Two student responses to the "Marci's Dots" algebra assessment item in the External Assessment System: (a) strategy involving a table; (b) attempt to use difference of differences.

with their answers. There is evidence that they were looking for some kind of pattern. They were on the right track. For example, this student recognized the pattern and explained it, but did not carry it out to the twentieth pattern number.

Many students attempted to solve this task using various approaches, and they showed evidence of applying what they studied in MiC units, methods that would not be considered in a traditional algebra program for eighth-grade students.

Connection to teaching and assessment practices. During instruction, Annette pressed students to visualize growing patterns from one figure to the next. She encouraged students to conjecture about the pattern, for example, in the eighth figure, and to use drawings to investigate the validity of the conjectures. Evidence of these practices was seen in student responses. In Similar Triangles, students identified the pattern and were able to extend the pattern through drawing. Some used algebraic expressions to generalize the pattern and subsequently used them in finding the number of triangles in the eighth figure. In the more challenging Marci's Dot task, many students attempted to find a solution, and some began to use a drawing strategy to visualize the pattern. Some students used algebraic notation in generalizing the pattern and methods learned when studying various MiC units. Their work was set out clearly, and their explanations were well formulated. Thus, Annette's efforts to have students do the mathematical work required in MiC units, visualize the next steps in each pattern, work toward generalizations, and write explanations of process rather than procedure were evident in student work.

Results of Problem Solving Assessments

The PSAs were more challenging for students in our sample. Although most students attempted all tasks, few responses were awarded full or partial credit (see Table 11.3). Annette commented that she perceived a lack of trying on the part of the students. She believed that students did not view these assessments as important because they were not state-mandated tests and the results were not going to be a part of their grade.

Seventh-grade PSA. Algebra items on the seventh-grade PSA were set in different contexts, each having two to five separate tasks. In the first context, Baby Feeding, the first task was a multiple-step problem that involved multiplication (see Figure 11.6). Three students solved this task correctly. For example, one student drew a line on the given graph to determine that a baby weighed 6.3 kilograms when he was 4 months old. Using this information, she read the appropriate data from the table, that 200 milliliters of milk

Table 11.3.   Results of algebra items from the
             Problem Solving Assessment System

	Full Credit*	Partial Credit	Attempt	Non-Response
Grade 7				
Baby Feeding				
Item 1	3	2	9	0
Item 2	0	2	12	0
Pyramids				
Item 1	3	0	10	1
Item 2	4	4	6	0
Item 3	5	—**	6	3
Item 4	2	—	10	2
Item 5	0	0	7	7
Playgrounds				
Item 1	8	—	5	1
Item 2	4	1	8	1
Item 3	0	4	9	1
Item 4	0	0	8	6
Grade 8				
Stretch				
Item 1	11	—	2	1
Item 2	0	0	8	6
Item 3	0	0	5	9
Item 4	3	—	5	6
Item 5	2	6	1	5
Cubes				
Item 1	0	4	6	4
Item 2	3	0	5	6
Item 3	0	1	8	5
Item 4	0	0	4	10

*Number of Students
**Partial Credit was unavailable in scoring

In this item from the Problem Solving Assessment System (Dekker, et al., 1997–1998), students connect pertinent information from graph with information in a table. They identify the appropriate arithmetic operation to calculate the answer.

Until they are about 6 months old, the only food babies eat is milk. One type of milk is made from a special kind of milk powder. This powder is called Nutri. Each box of Nutri contains a table that indicates how much Nutri a baby needs. This table is shown below.

weight of baby in kilograms	water in milliliters	+ spoonfuls of Nutri	is about milliliters of milk (each feeding)	number of feedings each day
< 3	60	2	65	7
3 - 3.5	90	3	100	6
3.5 - 4	120	4	135	5
4 - 5	150	5	165	5
5 - 6	180	6	200	5
> 6	180	6	200	4

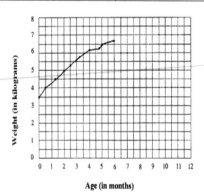

Figure 11.6.   Algebra assessment item from the Baby Feeding context in the Problem Solving Assessment System.

should be used for each of four feedings in one day, and she determined the correct amount of milk to feed the baby daily. Annette commented: "Very good. She even labeled each number. You can tell that this student is proud of her work." Most students looked for the weight on the graph and then went to the table, but some just wrote down the information, and one made a computation error. "I can see that most of them interpreted the graph and table well," Annette stated. "Only one student interchanged the information on the axes, and one student was overzealous. He showed the amount per feeding for everyone. He was hung up on unnecessary information, and he didn't go to the one necessary last step."

> For example: To prepare one feeding for a baby who weighs 2 kilograms you take 60 milliliters of water and add two spoonfuls of Nutri. This results in 65 milliliters of milk.
>
> Note: 24 hours = 1 day
>
> 1000 milliliters = 1 liter
>
> In the graph below the line shows how much Chris weighed during the first 6 months of his life.
>
> Use the graph above and the table on page 1 to find how many milliliters of milk Chris needed **per day** when he was 4 months old. Show your work.

The second multiple-step task involved substituting values into a formula, and the solution required conversion among metric units. Although most students attempted the task, none solved it correctly; conversion among metric units was problematic in responses that earned partial credit. After reviewing the student work on these tasks, Annette concluded, "I think everyone read these questions and actually thought about them." She commented that many responses showed that students read pertinent information from the graph and table, but stopped. She thought that the multiple steps required in the solutions were likely confusing for some students.

In another context, Pyramids, students responded to five algebra items set in a geometric context. In the first items, students were to identify the number of faces, vertices, and edges of given pyramids and complete a table to organize this information. Some responses were awarded full credit, although solutions were attempted by most students. Reflecting on student work, Annette commented:

> I'm not sure that they knew what face, vertex, and edge were. I'm not sure they could connect the terms and their meanings. Most of them

didn't seem to identify vertex, face, and edge. Maybe that's something I didn't emphasize enough with them.

In the next items, some students were able to determine numerical patterns in the table and extended the patterns for other pyramids. They then explained the pattern and wrote an expression for the number of faces, vertices, and edges in an n-gon. For example, one student stated: "All I do for the faces is add one to the number-gon [number of sides in the base]. On vertices I leave it the same as the faces, and for edges I x 2 the number-gon." Furthermore, some students used algebraic expressions in generalizing the pattern. For example, for the n-gon, given that the expression for the number of faces is $n + 1$, students wrote $n + 1$ for the number of vertices, and $n + n$ or n x 2 for the number of edges. In the next task, some students determined that Euler's formula does not work when $n = 5$. For example, "5 faces + 5 vertices − 7 edges = 3, so not it does not work for 5." Students were less successful in the last task, which involved combining like terms to make a concise formula for the n-gon.

In the final algebra context, Playgrounds, the last set of tasks on the assessment, students responded to four items. Many students were able to solve the first task, in which they identified the pattern in a sequence of growing playgrounds (see Figure 11.7) and found the number of gray tiles in the tenth pattern.

In the next task, some students correctly determined the number of white tiles in the tenth pattern and provided clear explanations of their work. For example, one student stated: "90 because 10 tiles in each row, then - 1 = 9. Then multiply that by 10." In the subsequent task, students were asked to write a direct formula for determining the number of white tiles given the playground length. No responses were awarded full credit, although some students wrote incomplete formulas, for example, "$n − n$ x $n =$" or "number of tiles x one less than the number." In the final task, determining the length of the largest playground that can be built with 1600 white tiles, none of the responses earned score points. For example, student responses included: "The contractor can use length 800 x 800 because 800 x 800 = 1600" and "1600 ÷ 2 = 800 gray, then 800 white." Annette noted the random operations: "I think there are some things students are not used to seeing in contexts. They are so used to just grabbing numbers and doing something to them. I think that was most evident in these responses."

Although students did not earn many score points in this context, Annette did see some positive elements in student responses: "There were a lot of attempts. Many recognized the pattern. Most did not name it correctly." However, Annette did not think the tasks were beyond the

A large **square** playground will be built in a park. It will be paved with gray and white square tiles in the following pattern: one diagonal will be paved with gray tiles, the rest of the playground will be paved with white tiles. The dimensions of the playground have not been decided. Some playgrounds of different sizes are shown below. The shaded tiles are the gray tiles.

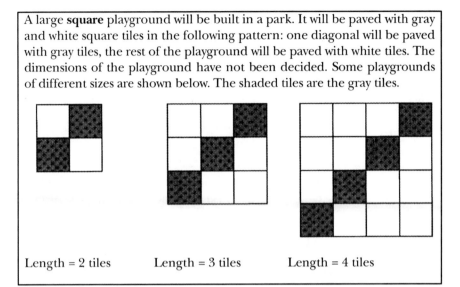

Length = 2 tiles   Length = 3 tiles   Length = 4 tiles

Figure 11.7.   Playgrounds context in the Problem Solving
                Assessment System (Dekker et al., 1997–1998).

reach of her students: "They knew there was a pattern, but didn't take the time to look for it and express it. I like the fact that a direct formula was requested. The last item on the assessment takes a long time to think about." Rather than concentrate on what students did not do, Annette looked for strategies that might lead to successful solutions and for particular skills that students demonstrated in their responses.

Eighth-grade PSA. Algebra items on the eighth-grade PSA were set in two contexts. In the first context, Stretch, students interpreted a table of data and a graph generated from an experiment involving a spring and different weights (see Figure 11.8). In the first task, most students predicted the length of the spring when an additional weight was used. In the next tasks, students were asked to calculate the slope of the line in the graph and write an equation for the line; all responses were incorrect. In subsequent tasks, students interpreted the meaning of the y-intercept and used points on the graph to make a comparison, but few responses were awarded full credit. Annette reflected on the lack of interpretation:

Some understanding is going on. Reading the graph is not the problem. Understanding what it means is. Slope was difficult for them. I guess I didn't teach that well. Most students know how to read what's

In science class, Nicola conducted an experiment involving a spring and different weights. The spring was stretched by hanging weights on a hook at the bottom of the spring. Nicola recorded the length of the spring after hanging each weight on the hook. The results of the experiment are shown in the table below.

Weight (g)	0	10	20	30	40	50	60	70	100	150	200
Length (mm)	300	305	310	316	325	329	332	...	354	380	406

Nicola used a computer to plot the results of her experiment on a coordinate system. This graph is shown below.

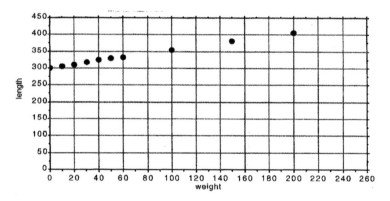

Figure 11.8.   Stretch context in the Problem Solving Assessment System (Dekker et al., 1997–1998).

plotted on that graph. That's good. That's a skill. But they don't know how to interpret. No one even seemed to understand that slope meant a change.

The second algebra context, Cubes, the last set of items on the assessment, was part of a series of PSA items designed to examine the development of algebraic reasoning over time. In the initial task, students were to identify the pattern in a sequence of growing cubes (see Figure 11.9), find the number of white cubes in the 4 x 4 x 4 cube, and explain how they determined the answer. Some responses were awarded partial credit. For example, one student responded: "48. If there is 12 on each side, 12 x 4 = 48," which implied that the large cube had only four faces and common cubes on adjacent faces were not considered. In the next task, students were given a formula for determining the number of white cubes in any large cube (number of white cubes = 12 x n – 16) and were asked to use it to calculate the number of white cubes in a 25 x 25 x 25 cube. Three students calculated the total correctly, substituting 25 for n and subtracting out 16. Another multiplied by 75, implying that n represented the product of 25 x 25 x 25, while others solely multiplied 25 x 25 x 25. These two responses again demonstrate the tendency for students in this sample to do something with the numbers rather than think about the task.

In the next task, students were to determine the number of black cubes in a 7 x 7 x 7 cube. One student drew one face of the cube, determined the number of black cubes on the face, and multiplied by six (accounting for the six faces). The student calculated the number of black cubes on a face incorrectly, but the strategy would have led to a correct solution. Another student realized that the number of black cubes on one face was 7^2, but neglected to account for six faces. Other students multiplied 7 x 7 x 7, while still others attempted to find numerical patterns. Few students responded to the last task; n^2 was the only response close to a formula for the number of black cubes in any large cube.

Annette had expected that students would do well on both the Playgrounds and Cubes contexts because "we modeled many problems in class that involved growing patterns." She felt that the three-dimensional situation compromised students' work:

> This was a little complicated for them. First of all, it was not clear to them that there were six faces on a cube, and they lacked the ability

The first three cubes of a larger sequence are shown below.

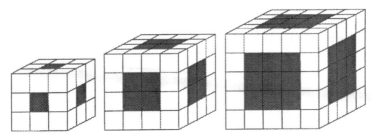

The first cube is a 3 x 3 x 3 cube. The second cube is a 4 x 4 x 4 cube. Small white cubes form the edges of the larger cubes. The other cubes on the outside are black.
Ignore the cubes on the inside.

Figure 11.9.   Cubes context in the Problem Solving Assessment System (Dekker et al., 1997–1998).

to recognize that the same small white cube was on more than one side. That was a weak area, maybe because of the three dimensions. Some students neglected to subtract out the 12 double-counted cubes. On the second item, some students substituted a value into the formula well. When looking at the two-dimensional playground [on the seventh-grade PSA], they didn't do too badly. The three-dimensional extension [on the eighth-grade PSA] was complicated for them.

The Playgrounds and Cubes contexts were the last items on the PSAs, and time may have been an issue for students. However, Annette perceived a lack of effort. She felt that students did not value these assessments because they were not going to affect their grades. At the same time, she was pleased to see that students did attempt the tasks:

> It appears that students understood the need to attempt any problem set before them. Though there were some students who did not respond to some questions, I am pleased that only a few students did this. Some students had some trouble giving accurate answers, but I value their attempts to organize the problem and answer it.

Connection to teaching and assessment practices. Annette's students had studied the eighth-grade MiC algebra unit in which slope was explored. Through her classroom assessment practice, Annette listened for and reacted to student misunderstandings. She adjusted lesson plans on the spot, as evidenced in the description of her teaching. In journal entries during the longitudinal study, Annette noted students' difficulties in understanding slope. For example:

> Students are not ready to move on in the unit. Their work on finding slopes is very poor. I insisted that we go over pages 20–22 again, and they will redo those pages for homework.

Lessons related to slope were reviewed and extended in an eighth-grade MiC statistics unit. However, Annette decided not to teach that section of the unit, and understanding of slope in a different context was not addressed again that year. In reflection, Annette said:

> As teachers, we look for mastery. We look for perfection. I have to remember that these are children. They are learning. So if they're picking up bits and pieces, their answers are a matter of what they did learn.

This statement is a reminder that understanding does not take place at one time, nor is growth in understanding a linear process (Hiebert & Carpenter, 1992). Rather, understanding develops in cycles of progression and regression. For learners, active reflective thought promotes integration of new ideas, forging rich connections that foster growth in understanding. In this case, revisiting and extending slope in an additional context was an important step in students' development. For teachers, understanding students' mathematical knowledge at particular points in time is essential in providing opportunities for students to reflect on the mathematics they are learning and to look for connections among mathematical ideas. Revisiting slope may have provided Annette with additional insight into students' reasoning, which in turn may have affected her instructional decisions, such as posing questions that linked lessons about slope in the algebra and statistics units.

In the Beams lesson and the Bricks lesson, as well as in other MiC lessons, students wrote direct formulas, substituted values in these formulas, and simplified formulas with multiple terms. PSA results indicate that students did attempt such tasks and suggest that students have not developed the necessary skills to solve them. Annette's encouragement of meaningful explanations was evident in student responses, as many students attempted explanations, even though their explanations varied in conveying critical points.

REFLECTION ON THE COLLABORATION

This collaboration gave Annette the opportunity to think about assessment results for her own students from research instruments, but Mary did not anticipate the degree of discomfort Annette experienced throughout this process. Annette wondered if she had been an effective teacher of mathematics—that her students understood concepts and had the ability to explain them well. She wondered how Mary and other educators who read about this collaboration would judge her effectiveness as a teacher. She commented, "I try so hard to affect gains in my learners' achievement that I fear their responses are evidence that I may have fallen short." As we analyzed the responses of students in our sample one by one, Annette was proud of some of the responses, but she wanted to know which of her students had done so well. She wanted to know whether students she expected to perform well did perform well and whether reluctant learners showed any progress. She pondered: "Was it possible that I was over- or underestimating my students' learning? I'm left wondering."

Annette struggled with the fear about what the results of the entire class indicated. But when we looked at the reports of the entire class, she was encouraged: "I finally felt relief when the reports revealed that success was attained overall. Some students clearly gained an understanding of generalizing patterns in algebra, while others did not." Furthermore, Annette felt that this collaboration was definitely worth the discomfort and fears. She learned how to constructively critique assessment results and felt it was a non-threatening way to share analysis of her students' work.

Mary witnessed a highly-qualified teacher coming to grips with assessment results from a sample of students who were frequently inadequate in terms of approaches to the tasks, communication of thinking, and reasonableness of solutions. In the finish, Annette chose to think about these results not as an affront to her teaching practices, but as an invitation to think about factors that might have influenced these results. It became apparent to Annette that these assessments did not have a high-stakes nature for students and that students did not put forth the effort necessary to develop reasoned responses. As the collaboration progressed, however, Annette began to think of students as learners of mathematics whose understanding develops over time. She did not make excuses for students. On the contrary, she looked for the skills that were discernable in student responses rather than what was lacking. Annette articulated the positive things she saw in student responses—the skills that were evident, the steps that were completed in multiple-step tasks, and the strategies that would have led to correct solutions.

Although this collaboration transpired several years after Annette's participation in the longitudinal study, she appreciated the opportunity to learn from her students' responses:

I was able to look at how each student answered the questions. I found it most interesting that most students demonstrated some knowledge of the questions. Many of their answers were incomplete, but I can see that they had some understanding of the problem. The students showed interest in attempting to answer the questions. The assessment showed me that many students were not as afraid of mathematics as they were when they began the school year in my class. The vocabulary was evident. The assessment encourages me to continue teaching using the hands-on, context approach in MiC. The assessment results are evidence that all students can learn. All students may not achieve a high level of success, but improvement will be evident.

Annette also thought about the things she wanted to emphasize with future classes. First, students need more work in completing multiple-step tasks. Second, students need more experience with manipulating three-dimensional objects and interpreting two-dimensional representations of those objects. Third, skills in algebra, such as the meaning and interpretation of slope and y-intercept, need further development during instruction in order for students to attain conceptual understanding. Finally, students need to work at developing thorough mathematical explanations.

As an outgrowth of the analysis of her teaching and assessment practices, Annette now includes more informal assessments throughout an instructional unit and insists that students use appropriate mathematics terminology during every class period. She is more convinced than ever that she must find ways to reach reluctant learners, the student who is turned off, or the one refusing to attempt tasks.

Through this collaboration, Mary realized the importance of this type of connection with study teachers. The collaboration was a unique opportunity for a study teacher to experience a different role on the research team—to look for factors that might have affected student performance over time. For Annette, the quality of her teaching practices was confirmed, and a way was opened for her to learn about how her students' work was viewed by those who scored it and to consider how that matched her own ideas. But the collaboration also enhanced Mary's analysis. In a very real sense, it confirmed the various aspects Mary considered in examining classroom instruction and students' opportunity to learn mathematics with understanding in the longitudinal study. It validated study findings that suggest that the most important aspects of instructional quality are the ways the lessons are presented and developed, the nature of the inquiry that transpires during instruction, and the evidence sought by the teacher as part of her classroom assessment practice. The collaboration allowed Mary to study, at a microscopic level, the relationship of these variables to assessment results through a teacher's consideration of the connections between classroom interaction and student performance.

The collaboration with Annette has also supported Mary's work outside the longitudinal study with both preservice and practicing teachers with respect to classroom assessment practice that is embedded in instruction. For example, during classroom interaction, what kinds of questions can be posed to generate discussion that will be fruitful for the teacher's assessment of mathematical understanding and at the same time promote learning for students? How can observational assessment and listening to students be beneficial in helping make future instructional decisions? What are effective ways to critique students' work so they can create more thorough answers? Attending to these points enhances the potential of a teacher's assessment practices to be grounded in interaction with students. Such assessment practice provides information that affects instruction by building on what students know and can do at particular points in time.

These methods promote teaching and classroom assessment practices that are well integrated.

IMPLICATIONS

We close this chapter with implications of our collaboration for teachers and educational researchers. Through reflection on her teaching, Annette has a deeper appreciation of the effect that her teaching and classroom assessment practices have on students' understanding of mathematics. The collaboration had a powerful impact on Annette's philosophy of mathematics teaching and assessment. Reviewing and critiquing one's teaching with respect to teaching for understanding, and analyzing responses on assessments designed for students to apply mathematics in new situations and reason and communicate mathematically, is an invaluable experience that promotes professional growth. We encourage teachers to use the descriptions in this chapter as vehicles to reflect on their teaching and assessment practices. What kinds of questions do you ask to gather evidence of student learning? What actions do you take in response to this evidence? How do your students respond to some of the assessment tasks presented in this chapter? How might your analysis of their work affect your teaching and assessment practices?

Educational researchers can consider providing the collaboration described in this chapter for teachers in their projects. Teachers can reflect on aspects of their own instructional practices before and in conjunction with professional development centered on classroom assessment practice. Teachers can then review the work of all students in their own classes on assessments designed for research purposes and look for linkages between instruction and the responses students generated. Preparation by the research team would involve selecting the content strand as the focus and preparing student responses to preserve anonymity. Discussion might focus on what the teacher notes about students' mathematical understanding and reasoning and potential approaches in future instruction. Moreover, discussion can include ways to assess students' future progress. For example, what tasks embedded in instructional units are useful for gathering information about a student's understanding of content and use of processes such as reasoning and communication? What questions might be posed as students work on problems or during class discussion to gather evidence of student learning and to summarize what students are learning? For more formal assessment occasions, what are the characteristics of high quality assessment tasks? How can assessment tasks be structured to get at higher levels of reasoning but still showcase skills? What methods can be used to help students write more thorough explanations that go beyond relaying steps in calculation? What methods can be used to think about changes in a student's work over time? This type of analysis and discussion has the potential for blending teaching and assessment practices and affecting students' growth in understanding mathematics.

ACKNOWLEDGMENT

The research reported in this paper was supported by the National Science Foundation REC-9553889 and REC-0087511, by the Wisconsin Center for Education Research, School of Education, University of Wisconsin–Madison, and Northern Illinois University. Any opinions, findings, or conclusions are those of the authors and do not necessarily reflect the views of the funding agencies.

REFERENCES

Carpenter, P., & Lehrer, R. (1999). Teaching and learning mathematics with understanding. In E. Fennema & T. A. Romberg (Eds.), *Classrooms that Promote Mathematical Understanding* (pp. 19-32). Mahwah, NJ: Erlbaum.

Cohen, D. K., McLaughlin, M. W., & Talbert, J. E. (Eds.). (1993). *Teaching for understanding: Challenges for policy and practice.* San Francisco, CA: Jossey-Bass.

Dekker, T., Querelle, N., van Reeuwijk, M., Wijers, M., Fejis, E., de Lange, J., et al. (1997-1998). *Problem solving assessment system.* Madison, WI: University of Wisconsin.

Fennema, E., & Romberg T. A., (Eds). (1999). *Classrooms that promote mathematical understanding.* Mahwah, NJ: Erlbaum.

Henningsen, M., & Stein, M. K. (1997). Mathematical tasks and student cognition: Classroom-based factors that support and inhibit high-level mathematical thinking and reasoning. *Journal for Research in Mathematics Education, 28,* 524-549.

Hiebert, J., & Carpenter, T. P. (1992). Learning and teaching with understanding. In D. A. Grouws (Ed.), *Handbook of research on mathematics teaching and learning* (pp. 65-77). New York: Macmillan.

Hiebert, J., Carpenter, T. P., Fennema, E., Fuson, K. C., Wearne, D., & Murray, H. (1997). *Making sense: Teaching and learning mathematics with understanding.* Portsmouth, NH: Heinemann.

International Association for the Evaluation of Educational Achievement. (1996). *TIMSS mathematics items: Released set for population 2 (seventh and eighth grades).* The Hague, The Netherlands: Author.

National Center for Education Statistics. (1992). *National assessment of educational progress: Grade 8 mathematics assessment.* Washington, DC: Author.

National Center for Research in Mathematical Sciences Education and Freudenthal Institute (Eds.). (1997-1998). *Mathematics in context.* Chicago: Encyclopedia Britannica.

National Council of Teachers of Mathematics. (1989). *Curriculum and evaluation standards for school mathematics.* Reston, VA: Author.

National Council of Teachers of Mathematics. (1991). *Professional standards for teaching mathematics.* Reston, VA: Author.

National Council of Teachers of Mathematics. (2000). *Principles and standards for school mathematics.* Reston, VA: Author.

Newmann, F. M., Secada W. G., & Wehlage, G. G. (1995). *A guide to authentic instruction and assessment: Vision, standards, and scoring.* Madison, WI: Wisconsin Center for Education Research.

Romberg, T. A., & Shafer, M. C. (Eds.) (2004). *Purpose, plans, goals, and conduct of the study* (Mathematics in Context Longitudinal/Cross-sectional Study Monograph No. 1). Madison, WI: University of Wisconsin, Wisconsin Center for Education Research.

Romberg, T. A., & Webb, D. C. (1997-1998). *External assessment system.* Madison, WI: University of Wisconsin.

Shafer, M. C., & Romberg, T. A. (1999). Assessment in classrooms that promote understanding of mathematics. In E. Fennema & T. A. Romberg (Eds.), *Classrooms that promote mathematical understanding* (pp. 159-194). Mahwah, NJ: Erlbaum.

Stigler, J. W., & Hiebert, J. (1999). *The teaching gap: Best ideas from the world's teachers for improving education in the classroom.* New York: The Free Press.

Wijers, M., Roodhardt, A., van Reeuwijk, M., Burrill, G., Cole, B. R., & Pligge, M. A. (1997). Building formulas. In National Center for Research in Mathematical Sciences Education and Freudenthal Institute (Eds.), *Mathematics in Context.* Chicago: Encyclopedia Britannica.

CHAPTER 12

TEACHING MATHEMATICAL DISCOURSE THROUGH CHARACTERS AND SCRIPTS

Marcia DeJesús-Rueff

Expeditionary Learning Outward Bound, Rochester, NY

INTRODUCTION

Let me start by giving a little background about myself. Now a middle-aged mother of three grown children, I have not always been a mathematics teacher. I taught for one year when I was twenty-three and then went into the business world, specifically market research. Sixteen years ago I left the corporate world and began teaching mathematics. At the time I conducted the research presented here, I was teaching grades six through eight in a large suburban school district in Western New York State. Among the courses I taught were sixth, seventh, and eighth grade sections of remedial mathematics; students were placed in these classes when they were not able to pass the regular grade-level course. For the past four years, I have taught high school in the same district and I now serve as the Mathematics Department Chair.[1]

Teachers Engaged in Research
Inquiry Into Mathematics Classrooms, Grades 6–8, pages 247–280
Copyright © 2006 by Information Age Publishing

The reason that this kind of research is important to me is that I have come to realize how little of what I teach actually takes root in my students' intellectual lives. Having spent so many years myself in the business world, however, I also realize that one's ability to think mathematically often leads to extremely interesting and creative work opportunities. My years of teaching have been spent looking for ways to foster mathematical thinking and reasoning with my students in order to provide them with the fullest possible future options.

The 1997–1998 School Year

The research presented here began eight years ago when I made a videotape of myself leading an eighth grade honors class to fulfill a requirement for a graduate course I was taking. In the videotape, students spent most of their time working in groups while I moved about from one group to the next answering questions and giving advice or prompts. Students were clearly engaged in their work; there was little off-task behavior. As the camera followed me around the classroom, everything looked great. To all appearances, this was a successful lesson focusing on collaborative learning.

Near the end of the class, however, as each group presented its findings, my belief in what my students had learned and in what I had taught them began to unravel. Students voiced several misconceptions, including basic misunderstandings of the difference between a variable and a constant; it became clear that they also had poorly formed understandings of the meaning of a negative slope. How could I have overlooked such basic errors while checking on each group? I had carefully monitored these groups. The students were not off-task; they were not playing around. Indeed, they gave all appearances of understanding the problems and of working together to solve them. Something important was missing in these groups. Simply telling students to work collaboratively and then monitoring their behavior had not automatically resulted in learning.

The 1998–1999 School Year

During the next school year, I focused my attention on answering this question: What kinds of teaching and learning strategies would help my students work through their misconceptions rather than simply accepting them as facts? I continued to have my students work in small groups, but I began studying these small group interactions through video and audio tapes of each small group. As I studied them, I saw that the groups in which individuals argued the most intensely with each other about the mathematical questions were the ones whose mathematical understandings grew most, as evidenced in the quality of their individual papers and tests. I found that these groups exhibited strong mathematical discourse as described by Cobb (1995) and Emerson (1996). In this type of discourse group participants actually build knowledge through: (a) creating a foundation for group procedures with dedication to the group, helping each other understand, actively listening to each other, and not giving in to one member's ideas because of her/his perceived "authority", (b) sound mathematical argumentation which includes backing, or proof, for all discussions, and a level of rigor appropriate to the grade of the students, and (c) intersubjectivity, or "taken-as-shared" meaning where the definitions, terms, and knowledge created through group discourse are recognized, used, and understood by all group members.

Through the video and audio tapes of these groups, I saw and heard the students challenging themselves to work through major mathematical misconceptions. My question for the next year became: How can a teacher actually teach students good mathematical discourse skills? I knew I could not simply lecture the students on what to do and then expect them to follow through. I toyed with the idea of showing them the video tapes from last year's classes, but I knew that would violate student confidentiality. I also knew that these video tapes would probably be quite boring to my eighth grade students because there is so much missing from the video tapes, including being able to see clearly the graphs the students are creating. Because several groups were video taped in the same classroom simultaneously, the sound quality of these videos was not good either.

The 1999–2000 School Year

I needed a vehicle to help students make major cognitive shifts in the way they interacted in collaborative groups. Having a student group argue in front of the class in a "fishbowl" format, where a small group holds its discussion in front of the rest of the class, could help the entire class see immediately what was going wrong in a given discussion; it could also run the risk, however, of embarrassing individual students. With help from my master's thesis advisor, I developed the idea of using plays to solve the problems of student confidentiality and embarrassment. Students would not be playing themselves; they would be playing a specific character with a given dialogue. That character would be the one making the mistakes. That character would be learning how to work better in collaborative groups. And that character could improve over time. This would also provide the opportunity for students to improvise, to deviate from the script in order to answer "What if" questions: What if this character had listened more? What if that character had not agreed to someone's ideas without discussing them?

THE RESEARCH STUDY

My overall question for this research was: Can characters and scripts help collaborative groups develop stronger mathematical group discourse in middle school classrooms? My hypothesis was that students who role play the important characteristics of mathematical discourse would develop skills necessary to be stronger members of collaborative groups in the mathematics classroom. Since their group skills were extremely weak, I decided to focus this research on my remedial eighth grade students. Once I saw the amount of time we had to invest in this process, I also had to limit this project to just the first four foundational aspects of good group discourse: helping each other understand, dedication to the group process, actively listening to each other, and not giving in to "perceived authority."

Through the characters and scripts, students had the opportunity to "try on" a variety of good and bad roles, as well as to watch other students do the same. They would determine which roles were the most successful and would begin using these "successful roles" in their own collaborative group discussions. These successful roles would thus become a part of each student's "inner dialogue"—their individual thoughts, self perceptions, and self-talk about their abilities to work with others in learning mathematics. By seeing themselves as active and successful group members, I expected they would work with their collaborative groups to construct new mathematical knowledge more efficiently. I also predicted that this stronger collaborative discourse would result in better individual understandings as well.

Additional questions I had included: (a) Would students respond to the idea of characters and scripts?, (b) How close could the students come to true mathematical argumentation as described by Cobb (1995) and Emerson (1996)?, and (c) Could the characters and scripts make a difference in how the students behaved within their small collaborative groups? In other words, would there be evidence of individual transformation? Evidence would include not only less off-task behavior, but also individual students' becoming more actively engaged in arguing mathematical concepts and ideas within a small group context.

I based the scripts on video tapes made during the previous two years. The discourse from those student groups provided numerous examples of both poor and excellent collaborative group discussions. Each of the elements necessary for good mathematical discourse (Cobb, 1995; Emerson, 1995) described above was found within these tapes. The video tapes also provided excellent situations where the mathematical discussions broke down; these became important features of the scripts, places where students could ask the "what if" questions to focus student thinking about their own discussions.

Overview of the 1999–2000 Study

Students practiced these scripts in small groups, then presented their play "fishbowl" style to the class as a whole. The entire class critiqued these presentations on how well the individual characters took part in their group's discussion. Gradually the class began asking "what if" questions about how certain changes could affect the group's discussion for better or worse.

At the same time, the real student groups worked through many of the same mathematical issues facing the characters in the small plays, namely the concepts of function, variables, and slope. As they held their own group discussions, I audio taped each group and later asked for a group and individual post-discussion analysis of how well the group was able to apply the principles of good group discourse. I also required students to write in their journals about the mathematical and group discourse concepts they were learning. Examples of journal prompts included: "Describe a time when you were in a small group that really listened to each other."; "What should you do when your group starts to get off task?"; "Write about a time when you felt a group really listened to your ideas."

As the groups progressed through these scripts, the entire class drew up a list of good group discourse to post in front of the room. As students added to this list each day, I provided additional structure for their statements by grouping their suggestions into more formal topics. Thus, "we should all work without talking about other things" was grouped under the header of "dedication to the group." Our classroom list became a bridge between the formal vocabulary of mathematical discourse research and the students' own words. This helped students remember what to do in their own small group discussions and also provided the framework for our post-discussion analyses.

Demographics

I taught these lessons in a large suburban middle school in Western New York State. The school has an enrollment of approximately 1300 students in grades six through eight. The student population is 95% Caucasian, and the majority of families fall within the upper-middle to upper class income brackets. This middle school does, however, also have an urban-suburban program which brings students from the nearby city, and it serves many lower-income students whose families live in subsidized apartment complexes, trailer parks, and farms within the district.

I was fortunate to have a teaching assistant in this particular class. She helped me in setting up the tape recorders, which ran at each group table, and she also provided an extra pair of hands for any additional assistance I needed, from finding masking tape to writing student comments on charts, to helping monitor group activity.

I completed the transcriptions and the analysis myself with assistance from my master's thesis advisor. Since this research formed the heart of my master's thesis, I had already budgeted the extra time into my evenings. The thesis structure also kept me going throughout this project: I met every other week with my advisor, prepared each section in a timely fashion to meet completion deadlines, received frequent feedback on my analysis, and also was required to prepare and present both a formal written thesis and a PowerPoint presentation to the Education Department's faculty.

The Focus of the Study

My students worked on the mathematical concepts of variables, slope and functions, a continuation of the work they had done with bottle graphs from the previous year. In that unit, students used bottles of different shapes and sizes and kept track of the height of the water in the bottle as they added measures of water. They then created graphs of the bottle graph functions. These bottle graph lessons on variables, functions, and slope came from a variety of materials, primarily those available from the National Council of Teachers of Mathematics.

The lessons and scripts I created for this study were based on extensions of this material. I collected a variety of unusual bottles and then made graphs for each bottle of the height of the water as a function of the number of cubic centimeters of water added. Another problem included creating a graph of their distance from bed in the morning for the first hour after their alarms went off. Yet another was to create a graph of the number of $10 shirts and $20 pants they could buy with a $100 gift certificate. The problems my students discussed in their own small groups were the same as those discussed by the characters in the scripts.

I believe the strength of these lessons lies in the way they begin with very concrete, hands-on experiences and then move into the more abstract concept of slope. The most important aspect of the lessons is that they require students to work together to build a shared understanding of slope. The teacher does very little direct teaching, relying instead on mini-lessons as needed for clarification about a specific skill or piece of information.

The lessons on group discourse required three days of 85-minute block class time. While this is a large amount of time, I believe it was well spent. Students were continuing to learn the mathematics in our curriculum, specifically developing an understanding of variables, functions and slope, while at the same time learning group discourse skills. Prior to this work, my students were often off task and not invested when they had to work in groups; afterwards, I found they were able to monitor themselves much better. I rarely had to step in to refocus small groups and, therefore, we were able to cover more material in more depth than we had prior to these lessons.

Scripts

I created eight small (five–ten minute) plays based on the information about group discourse errors gleaned from my previous videos. Two scripts, illustrating good and bad mathematical discourse, were devoted to each of the following topics: (a) dedication to the group, (b) helping other group members understand, (c) active listening, and (d) not giving in to authority (especially to those students with high social status). These social aspects of good mathematical discourse are based on the work of Cobb (1995) and Emerson (1996).

Each day, the script paralleled the students' own work. As I describe later in the outline of our daily lessons, the students worked on the same problem first, then read how these other "students" (the characters) had handled the problem in their own small group. My students had to determine which characters said which lines; I found this provided them with an opportunity to discuss each character's strengths and weaknesses in order to create a meaningful mini play, a play that actually fit each character as well as the discussion. Once the groups had assigned lines, they practiced the play together several times. Finally, one or two groups presented the play to the entire class. We then held class discussions about what had happened in the play and what this showed the students about the quality of learning. These ideas were listed on poster paper and kept on display in the classroom for the remainder of the year. The first script is shown in Figure 12.1.

We did this last year! It's dumb!

Not really, she pretty much left us alone as long as we were quiet.

But she was all over our butts when we sprinkled the water drops on each other....

Well, it did get a little crazy at times ... It was more of a mess than a few drops, remember. We did have a good time, though... And remember we got to go over to the high school!

Oh, yeah! That was OK. I wonder if we'll get to do that again this year...at least we'd get to miss science. That teacher really has it in for me – all he does is yell and yell just because I forgot a little assignment!

Or three!Or two!

And lunch!

Oh, man!

Did you see that new KORN video on MTV the other day? I thought it was really cool, the music was the best—I really liked the music and they had the best computerized effects —kind of like strobe lighting from the sixties but whacked out with all these bright neon colors and changing loca-tions – it was so cool!

Yeah, I liked it a lot but I still would rather watch hip hop—you know the kind of stuff on BET.

TEACHER: Two more minutes to finish your group's worksheet.

Guys, we'd better get our discussion going about what we remember about those bottle graphs ...we've only got a few more minutes!

Yeah, it was ok last year, but I don't want to do exactly the same stupid thing over again!

It doesn't seem to be exactly the same as last year. What's this stuff about role play and scripts and stuff?

Are we going to have to role play in front of the whole class?

What do you think—maybe we could videotape ourselves or something.

What is it we're supposed to be discussing anyway? I guess we'd better put something on this worksheet....Oh, geez, it's supposed to be a summary of our discussion! What am I supposed to write down???

TEACHER: Time for Our Whole Class Discussion!

Figure 12.1.   Script on Dedication to the Group

Note that this entire discussion shows a lack of dedication to the group. I designed this script to parallel many of the first discussions I audio taped for this research project.

Characters

Developing characters that would both attract middle school students and provide an avenue for teaching was a major consideration for this project. I finally based my four characters on historical personalities (Jack Kennedy and Amelia Earhart), a literary character (Nancy Drew), and an actor (Lou Diamond Phillips). I then fleshed out their personalities to highlight specific strengths and weaknesses each character would have within a collaborative mathematics group. My students remained unaware of the roots of these characters; interestingly, they never asked about where the characters came from.

The most important thing about these characters was that they had to become real to the students, both in their strengths and their foibles. Students needed to quickly understand each personality and had to be able to see each character's mathematical growth as the scripts progressed. The characters were also designed to provide the class with a way of discussing the elements of good/bad group discourse. For example, when critiquing their own groups, they might say that a group member had a difficult time arguing mathematically, "just like Amelia," or that another group member enjoyed being social but couldn't get down to work "just like Jack." The Personality Cards that I gave my students on the first day of the project are shown in Figure 12.2. I gave my students the following instructions: "Read the card for your character each time you play that person. This will help you show his or her true personality better as you role play."

NANCY: Nancy doesn't give up. She is a questioning person who is both clever and resourceful. Sometimes, however, she is outspoken. She is extroverted, occasionally moody, and she plays the trumpet. She doesn't avoid arguments and will keep talking until she wins. Because she is a natural leader, Nancy sometimes just expects that others will go along with her point of view. Nancy lives with her parents in a huge house with lots of land and woods all around it. Her friends love coming over to her house, but Nancy gets lonely when she's home by herself. She can't wait until she turns 16, because her father promised that if she keeps up her grades, he'll buy her a little car to tool around in.

AMELIA: Amelia is quiet, but very funny and brave. She has a naturally cheerful disposition, and she loves adventure as much as she loves reading about far away places. She is kind-hearted; her friends often tell her their troubles because they know she doesn't spread gossip. Unfortunately, often she will do anything to avoid arguments within her group. Amelia's parents are divorced; she lives with her mother in a small house near the center of town, but she often stays with her father, stepmother, and their two kids on weekends. She often leaves schoolwork at the wrong house and has to get her parents to bring her books, notebooks, projects, and gym clothes to school. She likes school, but sometimes she finds it hard to keep up and keep organized the way she'd like to.

JACK: Jack is quick-witted, athletic, adventurous, and outspoken. He is also wealthy and somewhat spoiled and self-centered. He is very friendly and loves working in groups but sometimes he has a hard time settling down to work. Jack's groups always have a good time; unfortunately, they don't always get much accomplished. Jack lives in an even bigger house than Nancy, but he has lots of brothers and sisters, both younger and older, to keep him company. People are always coming to stay at his house, and he always has interesting things to tell his friends about his parents' colleagues, since they both run an international business and must travel frequently. They have threatened Jack that if he doesn't pull his grades up, they may send him to private boarding school "where your teachers will watch you and tell you what to do 24 hours a day!"

LOU: Lou is quiet, resourceful, and kind of mysterious. Other students think he is "cool." No one knows too much about his background or his family. He came to our school a year ago and lives in an apartment with his older brother. Although never involved in trouble himself, he always seems to be nearby whenever other students are fighting or creating trouble, both in and out of school. He often hangs out with his older brother's friends. Lou is kind of detached when he works in groups. Talking over ideas in math class is hard for him, even though he almost always gets very high grades in math. He won't just go along with the group's ideas. Other students find it hard to work with him, because he doesn't have an opinion until he's heard all sides of an argument.

Figure 12.2.   Personality cards.

Overall Structure of the Lessons

Each of the three days' lessons progressed in a similar fashion. The teacher presented a mini-lesson and a "meaty" problem or topic for discussion. For example, on the first day students reviewed what they had learned the previous year. They were presented with several graphs and asked to describe the shape of a bottle that would create each graph. As a straight bottle is filled with water, the graph of its water height goes up a steady rate, creating a straight line. The height of the water in a bottle with a wide base that tapers toward a narrow neck will go up at an ever increasing rate, creating an upwardly curved graph. Another problem was to draw a graph of their own distance from bed as a function of time for the first hour after their alarm clocks went off. Other problems included matching graphs with stories (e.g., finding the graph of the height above ground level versus time of a child climbing a hill and then sledding down, finding the graph of the speed vs. time in the same situation). The final day's problem asked students to determine the number of $20 pants and $10 shirts they could purchase with $100 gift certificate.

The students worked as individuals for a few minutes then together in small collaborative groups to discuss the problem and to write up an explanation of their results. Each student wrote individually in his/her journal about what the group did and the results they came up with both as a group and individually. Groups received the first script, based on the topic they just finished discussing. They assigned each group member to portray a character and then assigned lines to the characters. They then practiced their play once or twice. The class came together and watched one or two of the groups present their plays in a "fishbowl" format. The whole class then discussed the play and listed on chart paper points of the discussion in which the characters did well and how they could have done better. The groups received a new play, an illustration of better discourse. Again they practiced the play, presented it to class, and we discussed why it worked better. I also encouraged students to discuss as a group and then as a whole class whether the new play did a realistic job and to explore how they themselves could work better within their own small group discussions.

Table 12.1 shows the topics for each day's work; the first two days were consecutive but the third lesson was held a week later to provide time for the first lessons to become more firmly set within our classroom culture.

Implementation

I carefully organized students into different groups each day. I did this intentionally, not wanting their familiarity with their group members to be

Table 12.1.   Topics for each day's lessons

Day	Lesson Topic	Script Topic
1	Review of Bottle Graphs	Dedication to the Group
2	Introduction to Slope	Helping Each Other
3	Reading Other Graph Clues	Active Listening
		Not Giving in to Authority
	Bottles to Graphs/ Graphs to Bottles	

the basis for improvement in mathematical discourse. On the first and third days, however, students were in the same group. This helped me determine whether their group skills had changed; in other words, could they work better with the same people in this group after learning some better group discourse skills? I judged the results by looking at examining their scores for each of the four elements of good mathematical discourse that we studied. Groups for a fourth day of the project, on which we did not work with scripts, were determined by my selecting students to examine more closely, based on their work during the three lessons. I grouped these four students together for this final day of audiotaping. This last day of the project took place about a month after the end of the formal lessons.

After the first group discussion on the first day, I passed out the Personality Cards to each student. What happened next was extraordinary. In every one of my classes, including the remedial class profiled here, complete silence descended upon the classroom; the students read the Personality Cards in their entirety, word for word. Most of my students, including those in the regular sections of eighth grade math, would not read a two or three sentence math problem without becoming bored and confused, yet these same students gravitated toward specific characters immediately, sometimes arguing about who should play which character. I stated to the class that sometimes girls would have to play male characters and vice versa, since the student groups were not always evenly split by gender. I expected this to create some difficulties within the groups, but it did not. Indeed, sometimes boys and girls reversed roles intentionally, seeming to revel in the opportunity to reverse genders.

When the students had finished reading the Personality Cards, I handed out the scripts, scripts based on the same mathematical problems the students had just been discussing in their own small groups. These scripts were unusual, however, because the lines had not yet been assigned to characters; instead, the students had to determine which character would say which line. After an initial few minutes of confusion and my explanation to give the lines to the character "to make the play make sense," the students jumped into the task wholeheartedly. Again, there was almost no off-task behavior as the students assigned the lines to the characters and practiced the play.

The first play on each element of mathematical discourse highlighted poor discourse skills; the second one "fixed" that mistake. Thus, students were exposed to contrasts between effective and ineffective mathematical discourse each day. Throughout this process, I video taped whole class work, audio taped individual collaborative groups, and collected student journals.

After they practiced, the groups took turns presenting the play in front of the entire class, "fishbowl" style. Most groups greatly enjoyed performing in front of their peers. Although I initially thought this would be the most important aspect in helping students recognize good versus poor mathematical discourse, I was wrong. Instead, what became the transforming elements of this process were the characters themselves.

After each "fishbowl" presentation, we discussed what had happened in the play. Students readily observed good and poor behaviors in the characters; these were discussed and written down on chart paper. At the end of each discussion, I labeled the element of mathematical discourse presented by the script (i.e., dedication to the group). As the three days passed, the scripts covered all the social elements of good mathematical discourse—dedication to the group, helping each other understand, active listening, and not giving in to authority. These same behaviors were repeated during each of the three days we worked on the scripts. Students were highly motivated, actively discussed which lines "belonged" to which character and why, and they greatly enjoyed themselves while at the same time learning to label specific kinds of group behaviors that help or hinder learning.

Analysis

My final analysis was designed to track the progress of four eighth-grade students— Jamal, Viola, Rob, and Carol (all names are pseudonyms)—as they learned about the first four elements of good mathematical discourse, those elements described above as creating a foundation for group procedures. These four elements provide the necessary basic elements which must be in place within small groups for students to be able to rigorously discuss mathematical concepts. I chose these four students because prior to this study, their group discourse skills were extremely weak. I wanted to know if the lessons on group discourse would help these students become better group participants. Information from seventh-grade standardized testing plus eighth-grade report card information high-light the students' abilities and academic achievement thus far. Back-ground information summarizes each student's family life, their socio-economic status within our school district, and their social adjustment to middle school. Selected quotes from each student's group discussions and written journal work provide a quick overview of their daily progress in learning and using good mathematical discourse. I also rated students' daily group work from all sources—videotapes, audiotapes, and journal entries—for each of the four elements studied on a simple rubric, with scores ranging from N for never to A for always.

Finally, I provide my own analysis of each student's growth during the month we studied these four elements of good mathematical discourse. This analysis looks at what happened in small group discussions and attempts to answer the question of why it happened. To do this, I put together individual student progress tables by using the transcripts of all the small group discussions these students were involved in, their journal entries, and my own notes from each day's work. Although subjective, I consistently tied my interpretations to the data gleaned from audio tapes of group discussions and from the student's own written journals in addi-tion to my personal observations as their teacher. I then put highlights of each student's work into a table which allowed me to track their progress day by day.

An important note is that I had used small discussion groups through-out the year with all my eighth-grade classes. Therefore, I cannot attribute the growth demonstrated by each student merely to their getting used to the small group format. Something very important happened in March, however, something that had not taken place in the previous six months of this school year: Within a few weeks, the students exhibited significant gains in their abilities to hold stronger mathematical discussions. I believe these gains are directly attributable to the characters and scripts I used to teach the first four elements of strong mathematical discourse.

Student-by-Student Analysis

Each of the four students I chose to analyze had been placed in a remedial mathematics program; all were in the same class. This placement was made through a combination of test scores, grades in previous mathemat-ics classes, and teacher recommendations. Most students in this class had struggled in or even failed regular mathematics classes in the past. Few knew their basic arithmetic facts or algorithms; all of them had to work very hard to use mathematics in real-life contexts.

It is important to note here the differences between these students and the average student in our district. On the whole in our district, our students score above the 85th percentile on national standardized tests, such as the Stanford Achievement Test. The average score on the Otis Lennon Intelligence Test is in the range of 110 – 115. Therefore, although some students in my study rank well within national averages on these tests, they remain significantly below the averages for our school district.

Jamal. Jamal's test scores were significantly lower than the district averages (see Figure 12.3). He is in the lowest quartile in his reading scores; his mathematics scores lag even below his reading scores. These combined with a below average I.Q. suggest that Jamal is at risk for failing the important new Regents exams which would mean he might not be able to graduate from high school. Despite these seriously low scores, however, Jamal is a very likeable, seemingly capable young man. He has tremendous social abilities and is sometimes very capable at solving mathematical problems.

Jamal is from an intact family and is a part of our Urban/Suburban program; he lives in the nearby city and is bused to our school. Although his economic status is low compared with other Bay Trail students, he is very popular and has many friends. He is a high-energy student who has difficulty

Jamal

Academic Background

Grade 7 Testing (all National Percentiles)

Stanford Achievement Test (SAT):	Total Reading	25	Total Math	14
	Comprehension	26	Problem Solving	18
	Vocabulary	18	Process	11
Otis Lennon Intelligence Test DIQ:	Total	85		
	Verbal	88		
	Nonverbal	85		

8th **Grade Report Card Averages:** Mix of C's and D's with an A- in music.

Figure 12.3. Jamal's Profile

remaining still or quiet for more than five minutes at a time. Despite his low test scores, he has considerable academic ability; however, he enjoys being social much more than paying attention in class. He is often in trouble academically; his grades hover just above the failing mark. He has not been a part of our Learning Center because he tended to disrupt the activities of others there with his incessant talking. This is Jamal's second year in my remedial mathematics class.

Jamal loves music and knows the words to all the current songs. He plays basketball and is often a part of social life at Bay Trail, even though it is difficult for him to get rides home from social events. He occasionally receives a disciplinary referral, primarily for showing off in class or for inappropriate actions in the hallways.

Selections from Jamal's discourse during this study are highlighted in Figure 12.4. Included are a few quotes and the daily discussion ratings (see Figure 12.5) which highlight Jamal's process during the four days of taped group discussions, days which were actually spread throughout the month of March. Day One's group discussions were typical of Jamal' group skills. He was easily distracted, frequently disrupted the group process, and was not at all concerned about doing the work assigned; he was particularly strong, however, in not giving in to authority. Jamal was, however, very enthusiastic about the scripts and about the characters, as evidenced by his quotes from the first day. He quickly identified with Jack, the character who is "quick-witted, athletic, adventurous, and outspoken." Indeed, Jamal played Jack in every single script.

Day One:

> "He's crazy – Look at him!"
>
> "Alright, we gotta make our graph, dude."
>
> Begins to explain his graph: "the distance from the bed…"
>
> Gets off track talking about his house alarm that goes off sometimes when he comes downstairs, his sister staying up late to watch movies, etc.
>
> "Don't you get it? It's easy!"
>
> Begins talking again about his graph and then immediately launches into an explanation about having had French toast sticks for breakfast.
>
> As soon as the teaching assistant leaves, James begins singing into the microphone and tapping on it.

Day Two:

"Then the second one (reading): 'They climbed the hill and then sledded down it.' 'Sledded down it – that doesn't sound right. Oh, well."

"I picked B because it looks like a little slopey hill thing."

Day Three:

"I thought the heart went with the red line because it can fill up faster at the bottom it is narrow and it get wider at the top. I thought the flask was the blue line because it wider at the bottom and takes longer to fill up. And the top is skinny and can fill up quick." (A written explanation of which graph goes with which bottle.)

Day Four:

"I'm dedicating. I have a lot of authority, too." (Joking around in his small group discussion.)

Figure 12.4.   Excerpts of Jamal's Discourse.

Table 12.2.   Jamal's Daily Discussion Ratings

	DAY ONE	DAY TWO	DAY THREE	DAY FOUR
DEDICATION	R	F	F	S
HELPING	R	F	S	S
LISTENING	N	S	S	S
AUTHORITY	F	F	F	F

Rating Scale: N = Never R = Rarely (less than 25% of the time) S = Sometimes (about 25 – 75% of the time) F = Frequently (more than 75% of the time) A = Always

Jamal's small group discussion skills peaked on the second day. He was able to acknowledge Erica's contribution to the group effort, as evidenced by the quote from Day Two above. This is the first time I ever heard him giving credit to someone else. He was highly dedicated to the group, consistently worked to help others understand, was a much better active listener, and continued to be strong in not giving in to authority.

On the third and fourth days, which were separated from the original two by a period of ten and twenty days respectively, Jamal's small group discussion skills went down somewhat but were still considerably stronger than they had been when we began the class work on role plays. There were, additionally, some distractions on the fourth day since it was the school's "Hawaiian Day," and many students had worn Hawaiian clothing to school. Interestingly, although Jamal was intent on singing into the microphone and making jokes during the group's discussion, his mathematical discourse remained strong. He did get the work done; this was clear when the teaching assistant came over and Jamal showed her his work. Additionally, Jamal was much less mean than usual when he teased Rob. Ordinarily, Jamal made fun of Rob in class, often to get a laugh from the other classmates. During this group discussion, Jamal teases Rob, but the teasing is much kinder. Indeed, there are moments when they actually joke around together, and Jamal treats Rob as an equal. He also gives Carol credit for the work she completes and shares with the group, again a rarity in my experience with Jamal. Finally, he jumps at the chance to share his group's work in front of the entire class, and he does so without any kind of joking around.

Jamal chose to represent Jack, a character that fit his own personality closely. The scripts then gave him the opportunity to flesh out this character in his own way, because the group had to assign the lines to the characters. Although I had originally thought the lines about hip-hop and BET might fit the character Lou, Jamal latched onto them immediately because they verified his own experience. Jamal is a highly verbal person; he learns much better when he can talk. He is also, however, often unfocused, distracted, and disruptive. Clearly, his verbal abilities need to be prized and used in a structured way. The scripts provided Jamal with the structure he needed to learn to focus himself in a way that his teachers had not been able to do despite years of encouraging, lecturing and nagging him to focus on the work at hand. Jamal learned that he could have fun and still get the group's task done. He found that presenting the task to the entire class, whether it was the solution to a problem or a carefully rehearsed skit, made him a real leader. He finally began getting the attention of his peers in a way that was acceptable to his teachers.

Viola. Although Viola's test scores all rank near the national averages, she was well below the averages for our district (see Figure 12.5). In sixth

Viola

Academic Background
Grade 7 Testing (all National Percentiles)

Stanford Achievement Test (SAT):	Total Reading	60	Total Math	38
	Comprehension	66	Problem Solving	48
	Vocabulary	50	Process	26
Otis Lennon Intelligence Test DIQ:	Total	95		
	Verbal	95		
	Nonverbal	98		

8th Grade Report Card Averages: Primarily C's with an A in
   physical education.

WISC-III (from 6th grade)	Verbal	98
	Performance	115
	Full Scales	06

Figure 12.5.   Viola's Profile.

grade, she was classified as Learning Disabled and was placed in inclusion classes for the next two years, classes with both a regular subject-area teacher and special education support teacher. Viola was declassified at the end of 7th grade. In eighth grade she had a 504 plan, allowing her to receive additional time for testing and special access to the Learning Center, but no other special provisions.

Viola is from a divorced family. She moved to our district from Texas after her parents' divorce, during the summer after fifth grade. Her family is on the lower end of the socio-economic structure for our district. Viola is, however, very active at the middle school, both socially and athletically. She is an excellent softball player, and she seems to know just about everybody in the eighth grade. Viola is not a sound student academically. She turns in about half of her homework assignments and rarely studies for tests. Her overall demeanor suggests that she is not really interested in learning. This was the first year I had Viola in my remedial class. Her grades year were low the previous year even with special education support, and her teacher recommended her for the remedial program.

Viola is fun loving and kind, and she has many friends. She is on the fringe of the students who cause trouble in school, but she has never been involved in any disciplinary action herself. Prior to the role play, Viola's usual method of interacting with her group experience was to giggle at everyone's jokes, do very little work on her own, and then whine until her teachers bailed her out by showing her step-by-step what to do. This is clear in the first audio tape. Viola laughs at Jamal's silliness then whines to the teacher, "I don't get it." She had very little dedication to her group, rarely helped other group members, never used active listening techniques and almost always gave in to authority. She herself explained in the above quote that when she felt clueless in science class, she simply went along with other group members who seemed to know what they were doing.

On the first day we used the skits, Viola chose to be Nancy, apparently giving in to the group's norms of girls playing female characters; the next day she began playing Amelia, but quickly exchanged roles with Rob. Viola began playing Lou, a character who fit her personality much more closely. Lou is described as "quiet, resourceful, and kind of mysterious." "No one knows too much about his background or his family. He came to our school a year ago..." Viola herself came to our school from Texas and, although popular, she is not well known by the other students. She herself could well be considered mysterious. Like Lou, Viola also knows many of our school's troublemakers well, but she herself is never involved in disruptions in or out of class.

The key point of Lou's character that is very different from Viola's is that he "won't just go along with the group's ideas." This is what Viola learns from the scripts; she begins taking charge of her learning by asking the group for help and by carefully using active listening techniques to be sure she understands. Her skills appear to have developed consistently over the four days of role plays until by the end she is not giving in to authority at all; she is very strong in helping others and in active listening, and she is dedicated to the group process (see Figure 12.6). In the final group discussion, she actively brings Rob into the group's fold by asking him questions about the song he says he loves, she clarifies her own understanding of the work Carol is doing, and she even threatens Jamal with "We're going to take your mic" to get him to settle down. At the end of this project, Viola has emerged as a leader in keeping her group on task and working as a cohesive unit. Figure 9 shows Viola's daily discussion ratings.

Day One:

"Mrs. DeJesús, can you help us?" (During small group discussion.)

Day Two:

"No, that just means that it's going along faster."

"That's where it (the graph) goes down."

Day Three:

"I think the red line goes with the heart shaped bottle because it takes a shorter amount of time to fill up. I think that the blue line goes with the flask because it takes longer to fill up because it's bigger & wider at the bottom so it takes longer to fill up." (Journal description of the day's problem)

"More people were participating." (Explanation of why the second "Helping" script group was better.)

"Jamal, you're good in math." (Small group discussion as Jamal presents his explanation. Viola acknowledges Jamal emerging stronger academic position.)

"I felt listened to in this class because my group was asking me questions and wanted to know more and we all gave each other the respect that they gave me." (Journal on a time when you were in a small group that really listened. Viola gets recognition and begins to find her own growing self-confidence.)

"I was in my science group and I knew that someone knew what they were doing and I had no clue so I gave in to their answers. (Journal on when you were part of a group in which some members gave in to the authority of others.)

Day Four:

"We're going to take your mike...." (Threatening Jamal, who is goofing around during their small group discussion. This is the first time I have heard Viola stand up for her group by going against another student's actions.)

"Our group was dedicated to getting the work done. We helped each other get the answers. Carol helped us get on the right track. We were all the authority, we all told each other to get on track."

Figure 12.6.   Excerpts of Viola's Discourse

Table 12.3.   Viola's Daily Discussion Ratings

	DAY ONE	DAY TWO	DAY THREE	DAY FOUR
DEDICATION	R	S	S	S
HELPING	R	S	S	F
LISTENING	N	S	S	F
AUTHORITY	R	R	S	A

Rating Scale: N = Never R = Rarely (less than 25% of the time) S = Sometimes (about 25 – 75% of the time) F = Frequently (more than 75% of the time) A = Always

The scripts gave Viola the practice she needed to be able to use the skills of active listening to increase her own learning. She found a character she could identify with, yet one that pushed her into an area of social interaction that she was not yet comfortable with, asking questions of and even challenging her peers. The scripts forced her, as the character Lou, to stretch beyond herself; even after the scripts were over, Viola retained their lessons in group discourse. Her progress suggests that characters and scripts can have a prolonged effect even after a short initial exposure.

Rob. Rob's test scores show a student who is above the national average in reading ability but who is struggles in mathematics. His I.Q. is exactly average, yet it is below that of the average student in our school (see Figure 10). This is Rob's third year in my class. Like Carol, he was with me for extra help in sixth grade and has been a member of my class in both seventh and eighth. He does not go to the Learning Center, because he usually does not have trouble getting his assignments done. He has good grades in math most of the time, but occasionally he flounders. He definitely needs the support of a small class. Rob has difficulty focusing in class and requires the smaller classroom atmosphere to do well in math.

Rob is from a divorced family. His family is in the lower mid-range of our district's families, middle class by national standards. Rob is a very immature eighth grader who has strong academic abilities; unfortunately, he often has trouble paying attention in class, and his grades suffer for it. He just has not yet pulled himself together academically or emotionally. Rob is often overwhelmed by middle school. Larger boys frequently bully him, and he is terrified of physical threats. Rob has few friends and is rarely picked to be a part of groups or teams. He is not involved in any sports or music activities.

Rob

Academic Background

Grade 7 Testing (all National Percentiles)

Stanford Achievement Test (SAT):	Total Reading	68	Total Math	15
	Comprehension	88	Problem Solving	30
	Vocabulary	32	Process	4
Otis Lennon Intelligence Test DIQ:	Total	100		
	Verbal	105		
	Nonverbal	96		

8th Grade Report Card Averages: Consistently B's with a smattering of
   lower grades.

Figure 12.7.   Rob's Profile.

Since he is so often the target of jokes and cruel remarks, it was easy for
me to overlook Rob's own lack of interpersonal skills. Although he was
dedicated to the group, Rob was very poor in helping others to understand, in active listening, and especially in giving in to authority. We see
this in his initial choice of character; he is simply given the leftovers by his
group both the first and the beginning of the second day (see Figure 12.8).
The first day he plays Amelia because the other group members have taken
the male characters. Rob goofs around, following Jamal's lead, and
announces, "I'm Amelia!" On the second day, he is assigned to Lou, since
the other boy in his group has already taken the role of Jack. Rob soon
decides that this role is not right for him and reads the personality card for
Amelia, who much more closely resembles him. He switches roles with
Viola, and from this point on Rob portrays Amelia as "Emilio," and Viola
continues to portray Lou. He states that he himself is also quiet but very
funny. Additionally, Amelia is from a divorced family, lives in a small house
with her mother, and often stays with her father and stepmother. This situation almost exactly parallels Rob's own. She is also a bit disorganized
because of her dual living arrangements and, like Rob, sometimes leaves
with the wrong schoolbooks, gym clothes, etc. Rob feels validated that he
is not the only person in such a difficult situation.

Day One:

"I'm Amelia!" (Copying Jamal, exactly.)

"Could that line be me?" (small group discussion when handing out lines for the different characters)

Day Two:

"Give a speech about feeling good and staying out of trouble if we do the work / work on it myself – *I'll* get all the credit." (Journal on what to do when your group starts to get off task.)

"I'm quiet, and I'm funny. I'm not very brave, though. I have a cheerful disposition. My parents are divorced. I fit her description perfectly!" (Reading and commenting on the character Amelia.)

"Because they are different bottles in how much water they can hold."

Day Three:

"OK, you can be Jacqueline, and I'll be Emilio." (To the female student playing Jack during a small group discussion)

Blue = Flask

Red = Heart bottle because the Blue starts out slow and ends up filling fast = (picture of graph), The red line starts up fast and ends up slow =(picture of graph) , like a heart, or an upside down flask.

"I think I'm going to do cartwheels." (Small group discussion)

"It's a noise machine!" (Talking about the tape recorder during a small group discussion)

"I don't think people have ever listened to my ideas – then again I usually don't have anything to say." (Journal writing about a time when you felt a group really listened to your ideas.)

"They're not really doing any work, they're just assuming."

"They're just assuming that all the graphs are straight lines."

Day Four:

"That's exactly what we're not supposed to do." (In response to Jamal' kidding around about giving in to authority.)

"I think we did pretty good. I said everybody participated. We stayed on task most of the time and we agreed but we didn't let authority take over." (Small group discussion of how well they followed the four components of good mathematical discourse)

"OK, you can stop now so we can continue our work." (To Jamal, asking him to stop playing around so the small group can continue its discussion; evidence of constructive assertiveness.)

"It goes in a constant pattern. I think it will be F (he selects a story from the list) because it goes in a straight line. And it's the only one in a straight line."

"Which way are we going to do the line? Are we going to do the line this way?"

Figure 12.8.   Excerpts of Rob's Discourse.

Indeed, the only quality Rob feels does not fit him from Amelia's personality is that "she is brave; I'm not very brave." It seems as though these scripts gave Rob permission to try being a bit braver. Over the course of Days Two and Three, Rob's group skills get a bit stronger, but by Day Four, he is able to completely reverse his usual tendency to give in completely to authority. On this day, he even challenges Jamal, who is goofing around, with "OK, you can stop now so we can continue our work." He is also listening better and helping others in his group understand. Table 12.4 shows Rob's daily discussion ratings

Rob is able to join in the fun of the group without the kind of harsh teasing he usually gets from Jamal. He sings, talks about a particular song he loves, and chats easily with Viola who asks him whether he owns the CD. He is being treated as a real member of their small discussion group, even though he does get silly during this final discussion in a way very much the same as earlier. Rob has changed from being acquiescent to saying what he really thinks, not just what he thinks others want him to say. Part of this results from practicing the scripts, but part of it arises fromhis group's accepting him, too.

Carol. Carol's I.Q. was at about average nationally, yet she is lower than average for our district. She consistently scored in the lowest quartile on national achievement tests, very significantly lower than our school's typical scores that rank in the top quartile (see Figure 12.9).

Carol's family is in the lower socio-economic segment of our district, lower middle class for the nation as a whole. Her parents are divorced, and her father was diagnosed three months before this project with inoperable cancer; he died the following summer. Carol is a very quiet and shy student who nonetheless has several close friends. She is a struggling student, and her progress is sometimes impeded because of her shyness. During the previous year, her other teachers considered having her tested for special education because she was lagging behind academically. Carol had me

Carol

Academic Background

Grade 7 Testing (all National Percentiles)

Stanford Achievement Test (SAT):	Total Reading	26	Total Math	24
	Comprehension	24	Problem Solving	36
	Vocabulary	38	Process	15
Otis Lennon Intelligence Test DIQ:	Total	98		
	Verbal	105		
	Nonverbal	92		

8th Grade Report Card Averages: Primarily B's & C's with a few lower grades.

Figure 12.9.   Carol's Profile.

Table 12.4.   Rob's Daily Discussion Ratings

	DAY ONE	DAY TWO	DAY THREE	DAY FOUR
DEDICATION	S	S	S	S
HELPING	R	S	R	S
LISTENING	R	R	R	S
AUTHORITY	N	S	S	F

Rating Scale: N = Never   R = Rarely (less than 25% of the time)   S = Sometimes (about 25 – 75% of the time)   F = Frequently (more than 75% of the time) A = Always

as her teacher in sixth grade for a special extra help mathematics section and has been my student for two years in remedial Math 7 and 8. Carol usually does well in my math class. She consistently turns in all homework assignments and studies regularly. She also attends our school's Learning Center for extra help with homework assignments.

Carol was often left out of our school's mainstream social life. She and her friends were on the edge both socially and economically. A couple of her close friends were students in our alternative education program and sometimes got into more trouble than the average students in our school. Carol was never been in any type of trouble at Bay Trail, but she was also never a part of any of our school's many social, athletic, or musical activities.

In Day One's discussion, Carol is a meek, silent student. She rarely interacts with the other two group members at all (see Figure 12.10). By the final day, she is still much quieter than the other group members, but she is standing up for herself, helping others to understand, listening and reflecting out loud on others' ideas, and is strongly dedicated to her group. Jamal and Rob both credit her with getting a solution to the group's problem and look to her as the real leader of their group. Rob states, "Yeah, Carol's the authority." Jamal chimes in with, "She's quite an intelligent girl." Indeed, Carol is the one who is really looking for mathematical taken-as-shared meanings when she asks, for example, "Should we use m for money?"

Day One:

"And then I don't know." (To herself as she is asking the teacher for help with her graph.)

"You could remind them to get back on task and hope they will." (Journal on what to do if your small group starts to get off-task.)

Day Two:

I don't care." (About which character she wants to be. There are already two girls in this group. One girl picked Nancy and the other Amelia. Carol ends up playing Lou.)

"I think role play is a good way for learning because everybody can be involved." (Journal on what do you think of role play in order to learn about group dynamics?)

"Well, the one that steeps up like..."

Day Three:

"I think we need to practice out loud with the parts." (Getting her group back on task.)

"No, Lou has to say it here. It's on the paper." (To A.J. who has missed writing down one of the parts.)

"Yes, I think it does, because we can actually see what people think and you can act it out and you understand it better." (Journal on whether you think the scripts we read and acted out as a class will help you discuss mathematics better?)

"I think the pink line goes with the heart shaped container, because. At the bottom of the bottle it is skinny & the graph goes up fast, & then the graph starts to curve. I think the blue lines goes with the flask because at the bottom of the flask it is fat & the graph goes up slowly & the flask gets skinnier & the graph goes up faster."

Day Four:

"Not." (In response to Jamal' saying that she is his girlfriend.)

"Why?" (Asking for an explanation. Carol is seeking justification, something she has not done before.)

"Like this, you go over here for ten shirts and don't go up for zero pants." (Helping others in her group understand how to graph the coordinate pair; this problem had students determine how many $10 shirts and $20 pairs of pants they could purchase with $100.)

"Should we use *m* for money?" (Asking for group consensus.)

Figure 12.10.   Excerpts of Carol's Discourse

Carol changes the most dramatically of the four students in this study. The first day she is barely a part of the group; she rarely acknowledges the presence of the other two group members except to give in to their playing around. The only time she asserts herself is to say that she would like to play Nancy; since she was the only girl in that group, she could have her pick of the two female characters. Her passivity continues into the next day when, in a different group, she agrees to play Lou after the other two girls pick Nancy and Amelia; she listens, however, and asks a few questions and sometimes helps others understand the work. By the third day, Carol *insists* on playing Nancy in the last two scripts, even though she also agrees to read Amelia's lines.

Nancy is a character quite different from Carol. Nancy is from an intact, extremely wealthy family with an indulgent father. Carol's own family situation is quite removed from this; her father, who is divorced from her mother, is dying of cancer, and her family's economic situation is not strong. The role of Nancy may have provided the opportunity for Carol to take control of her life in a world that was crumbling around her.

Nancy may also represent a way for Carol to try on personality traits that she is not brave enough to do for herself. Nancy does not give up; this is a strong contrast to Carol's profession on the second day "I don't care." Nancy is a "questioning person who is both clever and resourceful." She is also sometimes outspoken. By taking on the personality of a character so different from herself, Carol can speak and practice the words that are so hard for her to say on her own. The scripts provided her with a safe way to practice being a more assertive person. By the fourth day Carol is able to simply state "Not" when Jamal suggests she is his girlfriend; previously she would probably have just let the remark pass. Just as importantly, she is able to ask for group consensus, explain her thinking to the other group members, and even ask them for clarification. She is still a quiet person, but she is no longer passive. Table 12.5 shows Carol's daily discussion ratings.

Summary of Analysis

The characters and scripts gave students the ability to try new ways of interacting with each other, along the guidelines of good mathematical discourse. Students were no longer locked into familiar patterns; they were forced to try something new, in the guise of a different persona. This provided a non-threatening way for students to experience how good mathematical discourse feels as opposed to examples of poor discourse.

The scripts also provided the students with a sense of self-esteem. Rob, Viola, and Jamal all picked characters relatively close to themselves in personality. Each of these similar personalities also had something new to teach the students. Jamal learned that he could have fun and still get the work done in a small group. Viola learned that she could assert herself and take personal responsibility for understanding what other group members were discussing. Rob learned that he could be brave enough to disagree with other group members and that he could be a real part of the small group discussions.

Carol's situation was a bit different. She chose to portray someone very different from herself, someone much more outgoing and personable. She also chose to place herself in a socio-economic situation the exact opposite of her own; in doing so, I believe she was able to transform herself beyond the reality of her own home situation, a reality that has become increasingly difficult this year. Carol found that she did not have to become a loud person to joke around a little in a small group or to be recognized as a real leader; she simply needed to learn to assert herself, help others understand her work, and listen carefully to them. She also progressed the farthest in promoting argumentation based on taken-as-shared meaning.

By "fishbowling" their skits in front of the entire class, the students were able to practice new identities and ways of interacting in an acceptable way. No one was ever teased or made fun of for their efforts, even when their reading was not smooth or they acted silly. In this way, students could pick and choose the pieces of each identity that they wanted; they were not forced by peer pressure into acting differently.

By putting on new personalities, the students figured out the best way to interact with each other. They applied these axioms as they negotiated with their groups in assigning lines appropriate for each character. Students were then able to begin transferring these negotiations into stronger mathematical argumentation. Viola, for example, made sure she understood what the others were doing by actively asking questions; she also kept the group on task. Jamal validated others while at the same time making sure he understood the work, and Rob asserted himself in asking Jamal to not give in to authority. Carol developed the farthest by beginning to ask for taken-as-shared meaning while working on writing an equation.

Several important concepts became apparent as I completed the analysis of this project. First of all, there were three reasons why I believe this project helped students become stronger in their mathematical discourse:

- The scripts made the rules of good mathematical discourse transparent to the students; they were no longer expected to simply guess what was expected of them. Indeed, we could even think of these rules as the axioms of good mathematical discourse.
- The characters were what helped the students negotiate these skills. Their roles provided built in protection against public embarrassment. If they could not handle the skills as their real selves, they could try using the skills as a character.
- The negotiations the students had to make with the elements of good mathematical discourse and the given characters' personalities were the same kinds of discussions and arguments they had to do for real mathematical discourse to take place.

Table 12.5.   Carol's Daily Discussion Ratings

	DAY ONE	DAY TWO	DAY THREE	DAY FOUR
DEDICATION	R	S	F	F
HELPING	N	S	S	F
LISTENING	N	S	S	F
AUTHORITY	S	S	F	F

Rating Scale: N = Never   R = Rarely (less than 25% of the time)   S = Sometimes (about 25 – 75% of the time)   F = Frequently (more than 75% of the time)   A = Always

Additionally, it did not seem to matter to these middle school students whether they were portraying a male or female character; instead they looked to the character attributes to guide them. Thus, Viola played Lou consistently on the second and third days, and Rob portrayed Amelia, renamed "Emilio." Both students felt drawn by the more important aspects of personality than by gender.

I had three other questions that this research helped to answer as well. The first was: Would students respond to the idea of characters and scripts? From the very first time I presented the characters and scripts, when the students were absolutely silent and still, it was clear they were completely engaged. They were able to read the personality profiles and scripts and they were thoroughly absorbed by the process. Moreover, the way they vied for certain roles and lines proved their dedication to this project. Joe, for example, begged for a line by stating: "Please, please let me say this line. I'll never ask you for anything else ever again!"

My second question was: How close would the students come to true mathematical argumentation? I wanted to find out how to get students conversing well in mathematics classes. By having students assimilate characters and act out scripts of mathematical discussions, I have shown that they can develop the basic skills they need to be able to communicate ideas mathematically. While I was not able to teach the group discourse skills of argumentation and taken-as-shared meaning, given my own time limitations, I believe that those skills, too, can be explicitly taught using creative and compelling methods, such as role play and scripts. My research illustrates how important it is to teach middle school students via social methods.

My third question was: Did the characters and scripts make a difference in how the students behaved within their small collaborative groups? In other words, was there evidence of individual transformation? Middle school students are hungry for opportunities to understand and work on their social position. Recall how strongly Jamal and Rob identified with specific characters, even when those characters crossed gender lines. Note, too, how Carol was able to take on a new identity of a leader, something that had eluded her during the previous two and a half years of my class. I found also that the overwhelming majority of my students read and reread the personality cards; some even asked for more characters. Students this age are eager to understand themselves in terms of other people. Clearly, students want to learn about themselves, the social order of our culture, and how they as individuals fit into the social order of the culture at large.

Characters and scripts provide an active, social way to get students to take an active part in their own learning within the social context of a small discussion group. They show students how to work in a group both positively and negatively. The lines in a script, for example, show them how to get a group back on track as well as the importance of doing it. They also clearly illustrate how to use active listening, a weak skill for most students. Through the characters and scripts the elements of good mathematical discourse were made explicit to the students. My motives were transparent: I wanted them to be better participants in small groups in order to learn more.

The scripts made the "rules of the game" transparent to these students. They also provided the students ways to use those rules for their own best advantage. The characters, however, were of paramount importance in this study. The students used them as protection in testing out new ways of interacting and, I would argue, even "grew into" the characters themselves. Rob became braver, Viola found her own voice, Jamal discovered he could be a strong math student, and Carol found the true leader within herself.

My original question was "What kinds of teaching and learning strategies would help my students work through their misconceptions rather than simply accepting them as facts?" My daily notes illustrate that my classes continued to ask questions and to challenge each other more in the months following these scripts than they had ever done before. I had known three of the four students highlighted in this study for over two years before introducing them to the scripts and characters. During that time, our classes often worked in groups. Once the students had actively learned about mathematical discourse from the scripts, their group work became more focused, and the students achieved better grades in mathematics. I have lost track of Carol, who moved shortly after finishing middle school. While it is impossible to state with certainty that the scripts and characters we studied together when they were in eighth grade helped them achieve in later grades, Rob, Jamal, and Viola all graduated from high school in June 2004.

IMPLICATIONS

I used these scripts for one more year with similar results before I moved to our high school. I am currently working on scripts to be used with *Contemporary Mathematics in Context* (Coxford et al., 1997-2001), the reform curriculum being implemented at our high school.

I learned three things from this study that continue to influence my teaching. First, for reform curricula to be successful, in order for our students to learn with understanding, our students must use good mathematical discourse in their small group work. Secondly, this kind of discourse does not just happen; it must be taught. To be learned well, however, we must teach mathematical discourse in the same active way that we expect students to learn mathematics; we cannot simply lecture them about how to behave during group work.

Finally, I learned that classroom research helps me become a stronger, more vibrant professional. Not only have I had the opportunity to share my work with other educators, I have also experienced a profound connection with my students as we learn together. Their experience of me is no longer that of the holder of all knowledge; instead, I have myself modeled the kind of life-long learning professed in our district's mission statement. Students see me differently as they observe me working on my own classroom research, and I myself feel more empowered. Not only do I use other's research to shape my instructional practices, I can create new knowledge about teaching, too.

I found that scripts and characters provide a different way of getting students actively engaged in learning the foundational elements of good mathematical discourse. The four basic elements that we examined as a class—helping each other understand, actively listening to each other, dedication to the group process, and not giving in to perceived authority—appeared to improve significantly. Did this improvement with the foundational elements of good mathematical discourse then lead them to greater construction of mathematical understandings and the ability to justify their answers more clearly? I believe this work helped, but I did not attempt to collect evidence to prove this. I simply ran out of time to devote to this project. I do believe, however, that if there were a way to continue with this idea of small plays and characters in the day-to-day practice of math teaching, we could build beyond the basic foundational elements and gradually help students become stronger at both group discourse and their individual abilities to justify their answers mathematically.

Since adolescence is a time when our students "try on" many different roles, creating roles for them in mathematics classes is simply a natural extension of what they are doing already. Giving students a chance to try on the different roles in learning mathematical discourse is undoubtedly just one of many potentially powerful methods to improve instruction, however. By creatively aligning our instructional methods with age appropriate teenage behavior and interests, we educators can discover and test new techniques that enliven our classrooms while inspiring our students to learn deeper, more meaningful mathematics.

These practices must be created and examined inside our own classrooms, not from outside educational researchers. While research helped me tremendously in considering different possibilities and in helping me make sense of my observations, I was the one who created the scripts, carefully gathered data, and analyzed the results of my work. Ultimately, only dedicated practitioners can follow through with students on a daily basis to examine their practices within the context of real classrooms, test possibilities with real students, and complete research within the heart of their own classrooms.

NOTE

1 .   Through June 2005 I taught at Penfield Central School District, Penfield NY.

REFERENCES

Cobb, P. (1995). Mathematical learning and small-group interaction: Four case studies. In P. Cobb & H. Bauersfeld (Eds.), *The emergence of mathematical meaning: Interaction in classroom cultures* (pp. 25–129). Hillsdale, NJ: Lawrence Erlbaum.

Coxford, A. F., Fey, J.T., Hirsch, C. R., Schoen, H. L., Burrill, G., Hart, E. W., Keller, B. A., & Watkins, A. E. (1997-2001). *Contemporary mathematics in context: A unified approach* (Courses 1–4). Columbus, OH: Glencoe/McGraw-Hill, Columbus.

Emerson, A. W. (1996). Gender discourse in small learning groups of college-level developmental mathematics students. (Doctoral dissertation, Western Michigan University, 1996).

Breinigsville, PA USA
11 May 2010
237807BV00003B/30/A

9 781593 114992